MEMBRANES, CHANNELS, AND NOISE

MEMBRANES, CHANNELS, AND NOISE

Edited by

Robert S. Eisenberg

Rush Medical College
Chicago, Illinois

Martin Frank

National Institutes of Health
Bethesda, Maryland

and

Charles F. Stevens

Yale University School of Medicine
New Haven, Connecticut

PLENUM PRESS • NEW YORK AND LONDON

Library of Congress Cataloging in Publication Data

Workshop on Noise Measurements as a Probe of Ionic Conductance (1981: Denver, Colo.)
   Membranes, channels, and noise.

"Based on a Workshop on Noise Measurements as a Probe of Ionic Conductance held
February 25, 1981, at the Biophysics Annual Meeting, in Denver, Colorado"—T.p. verso.
   Includes bibliographical references and index.
   1. Membranes (Biology)—Electric properties—Congresses. 2. Ion channels—Congresses. 3.
Electric noise—Measurement—Congresses. I. Eisenberg, Robert S. II. Frank, Martin,
1947–   . III. Stevens, Charles F., 1934–   . IV. Title. [DNLM: 1. Membrane Potentials
—congresses. 2. Ion Channels—physiology—congresses. 3. Epithelium—congresses.
QH 601 W926m 1981]
QH601.W67   1981   574.87′5   84-15083

   ISBN-13: 978-1-4684-4852-8   e-ISBN-13: 978-1-4684-4850-4
   DOI: 10.1007/978-1-4684-4850-4

Based on a Workshop on Noise Measurements as a Probe of Ionic Conductance,
held February 25, 1981, at the Biophysics Annual Meeting, in Denver, Colorado

© 1984 Plenum Press, New York

Softcover reprint of the hardcover 1st edition 1984

A Division of Plenum Publishing Corporation
233 Spring Street, New York, N.Y. 10013

PREFACE

This volume is a collection of papers designed to increase awareness and utilization of fluctuation theory for the description of ionic events at the membrane. The papers are revised and updated versions of presentations made at a workshop entitled "Noise Measurements as a Probe of Ionic Conductance." As a result of discussions held at that meeting, the participants were asked to provide selected chapters designed to provide detailed descriptions of the theory and its application to a number of membrane channels.

Fluctuation theory was developed initially to explain statistical fluctuations of ordinary physical quantities such as random collisions between gas molecules and walls. As knowledge of ionic pores has advanced, it has become apparent that randomized fluctuations could be utilized to characterize pore behavior in excitable and epithelial membranes. Because of the increased awareness of the applicability of fluctuation theory, the workshop participants were invited to contribute papers to this volume to provide them with an opportunity to teach others the essentials of noise measurements. The emphasis of this volume is on the practical steps which must be followed to make and interpret measurements of noise, both noise produced by natural fluctuations of the transport system, and noise which is the response to an applied stochastic signal. This collection of papers is meant to emphasize practical limitations as well as practical and theoretical advantages of such measurements.

The original concept of a workshop on noise analysis was presented by the members of the Physiology Study Section, Division of Research Grants, National Institutes of Health. The workshop was organized by a committee consisting of Robert Eisenberg, Charles Stevens and Martin Frank and was held on February 25, 1981 as part of the Annual Meeting of the Biophysical Society held in Denver, Colorado. The support provided by the Program Committee of the Biophysical Society is gratefully acknowledged. The sponsorship of Caryl and Alan Erickson of the Dagan Corporation, Minneapolis, Minnesota is also greatly appreciated.

Special thanks are extended to my wife, Cheryl, for her support during this project and to Mrs. Florence T. Turska and Mrs. Veronica Q. Heller for the preparation of the manuscript in its final form.

Martin Frank

CONTENTS

INFERENCES ABOUT MOLECULAR MECHANISMS THROUGH

FLUCTUATION ANALYSIS

Charles F. Stevens

Section of Molecular Neurobiology
Yale University School of Medicine
New Haven, CT 06510 USA

Membrane channels operate in an inherently probabilistic fashion because the world at a molecular level is chaotic. As a consequence of this molecular chaos, the number of open channels in an excitable cell's membrane varies incessantly around an average value and produces random fluctuations in membrane conductance. The statistical characteristics of these fluctuations are not identical for all channel types, but rather reflects those same physical processes that give rise to variety in behaviors between species of channels. Thus, one sort of channel may produce small and rapid fluctuations, whereas another may yield large, slow noise. The essence of fluctuation analysis --the study of a system's inherent noise--is this: channel noise reflects underlying molecular mechanisms, so we can learn about these mechanisms by studying conductance fluctuations. In some situations, fluctuation analysis is simply an alternative to studying the average response to a perturbation, but in other instances we can gain information not otherwise available. For example, different molecular mechanisms can exhibit identical average behavior, but produce fluctuations with distinct characteristics. Fluctuation analysis can therefore be used to

1

distinguish between various mechanisms underlying some macroscopic process.

The goal of this paper is to present, in a simplified but relatively complete manner, the essentials of fluctuation analysis. The main discussion divides into three sections. In the first section, I will deal with the characterization of noise. The section contains a description of spectral analysis and the covariance function (nearly the same as the <u>autocorrelation</u> function). The second section treats the connection between molecular properties and inherent fluctuations. I will present a specific, simple molecular mechanism for channel gating, derive a probabilistic description of this mechanism, and calculate the spectrum of conductance fluctuations anticipated. The final section considers certain practical considerations in fluctuation analysis including a discussion of some physical sources of noise, and sampling theorem, and aliasing.

NOISE IS CHARACTERIZED BY SPECTRAL ANALYSIS

Figs. 1B and 2B present two samples of noise--the samples might represent, for example, conductance fluctuations as a function of time--which are clearly different; the trace in Fig. 2B is obviously more rapidly varying than the one in Fig. 1B, although the amplitude of the fluctuations appear similar. Any useful characterization, then, must in some way specify both amplitude and rapidity. One usual way of characterizing such noise samples is by spectral analysis. The basic idea of spectral analysis is a simple and natural one: noise records are decomposed into their constituent sine and cosine wave components, and the amplitude and frequency of these components are used as the required specification of noise properties.

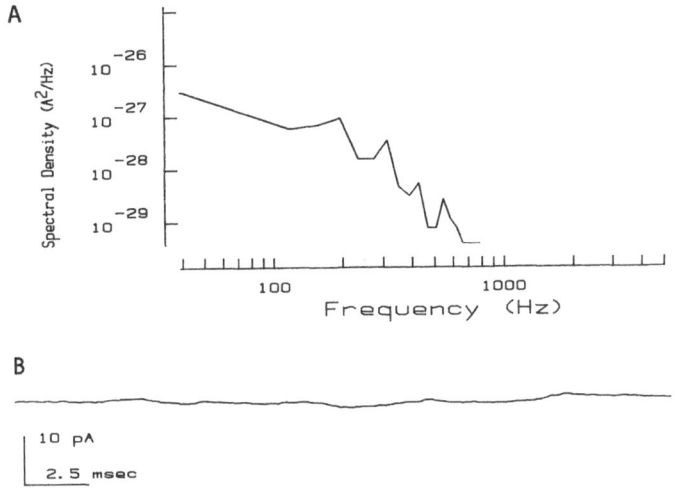

Fig. 1.   A.   Spectrum of noise samples (shown in B) presented in a
double logarithmic plot.
B.   Sample of relatively slowly varying noise.

The rapidity of sinusoidal wiggles is specified by a single
number, frequency, but two distinct measures of sinusoid size can
be used.   One size measure is the amplitude (one-half the peak to
peak excursion) and the other is the RMS (root mean square) value.
The RMS value is calculated by squaring the sinusoid (so that the
positive and negative values cannot cancel in the averaging
process) and finding the time average of this squared sinusoid.
The RMS value is then the square root of this time average.   Volt
meters normally read RMS values rather than actual amplitudes
because a sinusoid with a particular RMS value has the same power
as a DC voltage of that amplitude.   For example, the standard
U.S.A. household voltage has a frequency of 60 Hz, an amplitude of
about 160 volts, and an RMS value of about 110 volts with a power
equivalent to a 110 volt DC source.

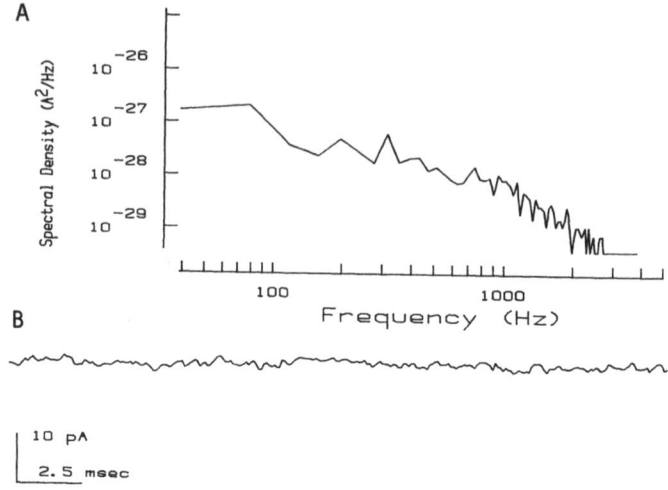

Fig. 2. A. Double logarithmic plot of spectral density vs.
         frequency. Spectrum calculated from relatively rapidly
         varying noise displayed in B.

The key to characterizing noise in terms of its constituent
sinusoid components is Fourier analysis. According to Fourier's
theorem any usual function of time can be decomposed into a sum of
sine and cosine waves with various amplitudes. Let us suppose
that we have a sample of a noise with zero mean, and further
suppose that the noise contains no frequency components above 500
Hz. The original noise sample (with a duration of T seconds and
amplitude y(t) at each time) can, according to Fourier's theorem,
be written as

$$y(t) = A_1 \cos\left(\frac{2\pi t}{T}\right) + B_1 \sin\left(\frac{2\pi t}{T}\right)$$

$$+ A_2 \cos\left(2 \cdot \frac{2\pi t}{T}\right) + B_2 \sin\left(2 \cdot \frac{2\pi t}{T}\right)$$

$$+ A_3 \cos\left(3 \cdot \frac{2\pi t}{T}\right) + B_3 \sin\left(3 \cdot \frac{2\pi t}{T}\right)$$

+ .....

$$+ A_k \cos \left(k \cdot \frac{2\pi t}{T} \right) + B_k \sin \left(k \cdot \frac{2\pi t}{T} \right)$$

+ .....

$$+ A_{500} \cos \left(500 \cdot \frac{2\pi t}{T} \right) + B_{500} \sin \left(500 \cdot \frac{2\pi t}{T} \right) \qquad (1)$$

The amplitude of A_k of each cosine wave and B_k of each sine wave is calculated according to the equation

$$A_k = \frac{2}{T} \int_0^T y(t) \cos \frac{k2\pi t}{T} dt \qquad (2a)$$

$$B_k = \frac{2}{T} \int_0^T y(t) \sin \frac{k2\pi t}{T} dt \qquad (2b)$$

Notice that, if T is one second, A_1 is the coefficient of the cosine with frequency 1/T Hz, etc.

Because the noise sample in our example contains no frequency components about 500 Hz, the 1,000 coefficients (500 A's and 500 B's) in front of the sinusoids displayed in equation 1 contain precisely the same information as the original noise record; that is, the noise record can be exactly reconstructed by adding together 500 sine and cosine waves with the amplitudes given by the A's and B's as shown by equation 1. We are interested in characterizing the amplitude and frequency components of the noise, but not in the details of the exact wave form of the noise; this varies from sample to sample even though the "general appearance" of the noise does not change. The appropriate way of giving a general description of the noise is to square the coefficients (so they cannot cancel--the various A_k and B_k can be either positive or negative) and average the coefficients for each frequency. That is, the spectral density, S_f, is defined for each frequency by the equation

$$S_f = (A_f^2 + B_f^2)/2 \qquad\qquad (3)$$

Here the subscript f indicates a coefficient for a component with
the frequency f. An important property of this measure is that,
although a particular physical source of fluctuations will give
rise to noise samples which always vary in precise waveform from
sample to sample, the estimates of spectral density will converge
when measured with sufficiently long samples to a unique
characterization of the noise source known as the spectrum. The
usual way to present such spectral densities is on a double
logarithmic plot so that a wide range of amplitudes and
frequencies can be conveniently displayed. Plots of log spectral
density vs. log frequency are shown in part A of Figs. 1 and 2 for
the noise samples shown in part B of those figures.

The two spectra presented in Figs. 1 and 2 have about the
same shape (both decrease at higher frequencies) but differ in the
frequencies represented (the fast noise has more high frequency
components than does the slow noise). Spectra do not always have
the same form as those illustrated in Figs. 1A and 2A; various
noise sources can have spectra which differ markedly in their
general shape. Some examples are shown in Fig. 3. We can gain
information about a system, then, from: the noise amplitude,
range of frequencies present in the noise, and shape of the
spectrum.

Just as exponential functions frequently arise in physical
relaxation processes, so does a particular spectral form, called
the simple spectrum, appear frequently in fluctuation analysis.
The simple spectrum, illustrated in Fig. 3A, has the equation

$$S(f) = \frac{2\sigma^2/\pi f_o}{1 + (f/f_o)^2} \qquad (4)$$

and is characterized by two parameters, σ^2 (the variance of the noise) and f_o (the corner or cutoff frequency). The simple spectrum approaches a constant spectral density $(2\sigma^2/\pi f_o)$ and in the high frequency range above the cutoff frequency approaches spectral densities which decrease like $1/f^2$. The corner frequency corresponds to the point (corner) at which these two asymptotes (the constant low frequency one and the high frequency $1/f^2$) intersect. Simple spectra are also called "$1/f^2$" (because of the behavior at high frequencies), "single time constant" (because their high frequency behavior is characterized by a single time constant parameter equal to $1/2\pi f_o$), "Lorentzian" (because equation 3 is similar to the Lorentzian function used in nuclear magnetic resonance studies).

The spectrum is only one of two commonly used ways for characterizing fluctuations: the other is the covariance function. The covariance function specifies over what time periods the noise is correlated. If fluctuations are slowly varying, as in Fig. 1B, knowing that $y(t)$ is, for example, positive (that is, above the mean) at a particular time t you have a good chance of being right if you predict that the noise will still be positive 2 ms later. If the fluctuations are rapidly varying, as in Fig. 2B, knowing that the noise is about 0 now is almost no help in predicting what it will be at 2 ms from now. The covariance function is defined as the average of y (now, at time t) times y (T seconds from now, at time T + t). Specifically

$$C(t) = \frac{1}{L} \int_o^L y(t)y(t + T)dt \qquad (5)$$

where L is the length of time over which the (time) average is

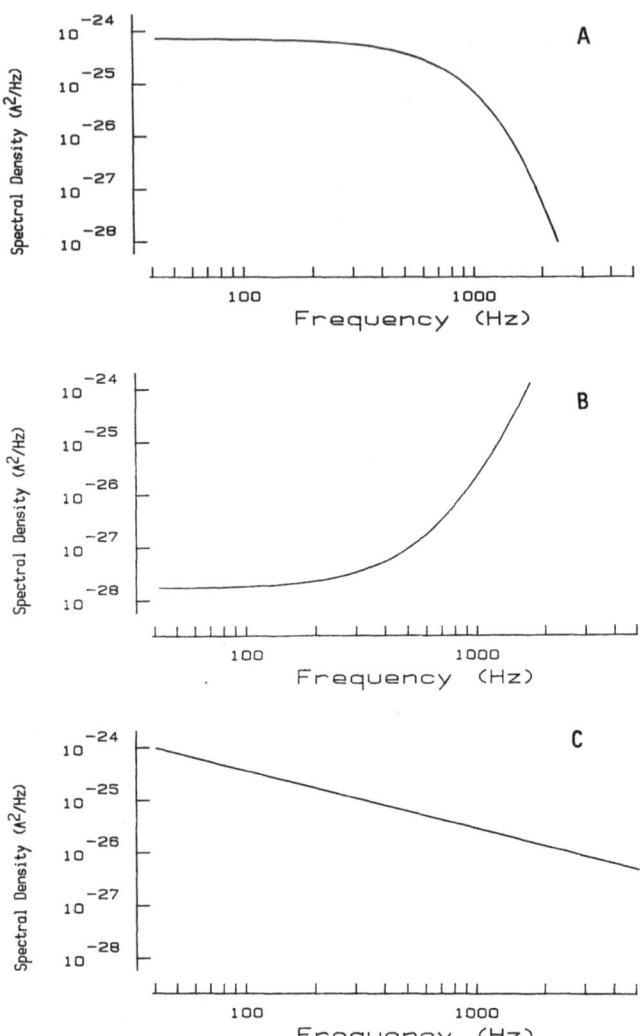

Fig. 3. Three common spectral forms.

A. Simple spectrum described in equation 4. This
spectrum can arise from a simple open-close channel
mechanism.

B. Spectrum with the equation $S(f) = A + Bf^2$. Spectral
with this form can arise from certain carrier transport
mechanisms.

calculated and the covariance function is denoted by $C(T)$. Notice that for $T = 0$, the covariance function is simply the time average squared amplitude of the fluctuations; that is, $C(0)$ is the variance. Simple molecular mechanisms give rise to exponential covariance functions:

$$C(t) = \sigma^2 e^{-(T/\tau)} \tag{6}$$

This covariance function is specified by two parameters, the variance σ^2 and characteristic time τ. The noise is uncorrelated for times large compared to the characteristic τ. Because the covariance function indicates the time range over which the noise is correlated with itself, it is sometimes causally called the autocorrelation function, although strictly speaking this term should be reserved for the covariance function normalized to unity at time $T = 0$.

Two useful characterizations of the same process must be related. As might be expected, then, the spectral density and covariance functions contain the same information: one can be calculated from the other by a Fourier transform. Those familiar with Fourier transforms will see at once that taking the Fourier transform of the convolution integral that appears in equation 5 will give the spectral densities as defined earlier. The spectrum that corresponds to the exponential autocorrelation function

C. Spectrum with the form $S(f) = A/F$. This spectrum describes which is called "one-over-f noise", a ubiquitous form of noise arising from electronic components and very many physical and other sources (e.g., stock market prices).

(equation 6) is the simple spectrum described by equation 4. The parameter σ^2 has the same meaning in both equations 4 and 6, and the time constant in the autocorrelation function (equation 6) is related to the cutoff frequency f_o in the simple spectrum (equation 4) by $\tau = 1/2\pi f_o$.

PROBABILITY THEORY GIVES THE LINK BETWEEN SPECTRUM AND MECHANISM

The preceding section has described two related ways for characterizing fluctuations, the covariance function and the spectrum. To gain information from fluctuations, we must find the link between physical characteristics of the system--gating properties of channel molecules, for example--and the spectrum (or equivalently covariance function). The goal of this section is to illustrate, by a simple example, the method of forming this link.

The summary of the procedure is as follows: we begin with a postulated molecular mechanism for channel gating. The mechanism is next formalized by deriving a differential equation which governs the probability of finding the channel in its various permitted states. This equation is then used to calculate the covariance function which is Fourier transformed to give the predicted spectrum. The predicted spectrum can in turn be used for comparison with experimental results to test the validity of the proposed mechanism, and once this mechanism is confirmed, to extract information about the system.

Suppose that a channel protein can exist in only two states, open and closed:

$$\text{closed} \underset{\beta}{\overset{\alpha}{\rightleftarrows}} \text{open} \tag{7}$$

The rate constant for the open to closed transitions will be called β, and α will be the opening rate constant. We suppose that these rate constants are fixed and that the channel is fluctuating between the open and closed states. First of all, we wish to find the differential equation that describes the probability of finding the channel in its open state, and then to find an expression for the covariance function and spectral densities for the fluctuations in a number of open channels.

We can discover immediately the steady state probability f that this channel is open. The net rate of transitions from open to closed is βf, where β is the closing rate; f is the fraction of channels open and $1 - f$ is then the fraction closed. The net opening rate correspondingly is $\alpha(1 - f)$. In the steady state net opening and closing rates are equal so that $\beta f = \alpha(1 - f)$ and f is therefore $\alpha/(\alpha + \beta)$. This gives us the steady state open probability.

We suppose, as is usual in physical applications of this sort, that the probability of a transition from, say, closed to open approaches Δt from small Δt. Let P(t) be the (joint) probability that a channel is open at time 0 and also open at time t. A channel can be found in the open state at time $t + \Delta t$ in only two ways (for Δt so short there is not time for more than 1 transition): either that channel was open at t and did not close during Δt or it was closed at t and opened. These two possibilities can be expressed in the equation

$$P(t + \Delta t) = P(t)(1 - \beta\Delta t) \quad + \quad ((1 - P(t))\alpha\Delta t$$

open <u>and</u> did   <u>or</u>   closed <u>and</u> then
not <u>close</u>   opened

When we rearrange this equation and take the limit as Δt approaches 0, a differential equation governing the joint probability of a channel being open at 0 and also open at t is

$$\frac{dP(t)}{dt} = -(\alpha + \beta)P(t) + \alpha \tag{8}$$

Before continuing with our derivation, a brief detour is
necessary to explain the difference between two kinds of
averaging, time and ensemble averages. In time averages we
perform the averaging process on one record over time whereas for
ensemble averages, the averaging process is at a fixed time across
a large ensemble of systems generating random records. To be
specific, suppose that w had one record specifying the state of a
single channel over time, and at each point in time the channel
could be either open or closed. The probability of this channel
being open would then be estimated by the fraction of time during
the record for which we observed the channel to be open. Imagine
now a large population of identical independent channels, each of
which can be either open or closed. The probability of an open
state is now estimated by the fraction of this population of
channels that we would find open at a particular time point. For
most systems with which we are concerned, these two averaging
procedures yield the same results. Such systems, for which time
and ensemble averaging can be interchanged, are called ergodic.

To calculate the average of a quantity, one always weights
each possible value the quantity can have by the probability of
that value and sums over all possibilities. For example, the
average grade on an exam for a class is the sum of all possible
grades, each weighted by the fraction of the class that received
that grade. For the average conductance of our two state channel,
then, we would find $g = \gamma f$ where g is the average conductance and
γ the conductance of the channel when it is open (channels are
assumed to have zero conductance when they are closed so the
contribution $0 \cdot (1 - f)$ of closed channels to the

average vanishes). This is an ensemble average because we are assuming that f is the fraction of channels in the open state found in the large ensemble of such channels.

The covariance function is, by definition, the average of the conductance (now) _times_ the conductance (T seconds later) _minus_ the mean conductance squared. Because the conductance of a closed channel is zero, all terms in this average for the scheme (7) vanish except the one weighted by the probability of being open at time zero _and_ open at time t. That is, the covariance is

$$C(t) = \gamma^2 P(t) - (\gamma f)^2 \tag{9}$$

The general definition of the covariance function for states with different conductance levels is a little more complicated. Let $P(i,j,t)$ be the joint probability that the system is in state i initially and in state j t seconds later. If $\gamma(i)$ is the conductance of the channel when it is in state i, the covariance function is given by the double summation

$$C(t) = \sum_{i,j} \gamma(i)\gamma(j)P(i,j,t) - \bar{\gamma}^2 \tag{10}$$

where the subscripts i and j range over all possible states the system can attain and $\bar{\gamma}$ is the average conductance. For our simple two state channel, each f these indices could take on a value corresponding to either open or closed so the sum would contain four terms.

It is clear from equation 10 that we can calculate the covariance function, and then the spectrum, if we know joint probability that a channel is open at time 0 and also at time T as specified by equation 8 for scheme 7. This differential equation is easily solved with the initial condition that $P(0) = f$; the probability that a channel is open at time 0 is the steady state open probability f. The solution is

$$P(t) = fe^{-(\alpha + \beta)t} \tag{11}$$

The covariance function is then (according to equation 9)

$$C(t) = \gamma^2 f(1 - f)e^{-(\alpha + \beta)t} \qquad (t>0)$$

with   (12)

$$f = \alpha/(\alpha + \beta)$$

Our simply molecular scheme, therefore, predicts an exponential autocorrelation function which, as indicated earlier, corresponds to a simple spectrum of the form

$$S(f) = \frac{2\gamma^2 f(1 - f)/(\alpha + \beta)}{1 + (2\pi f/(\alpha + \beta))^2} \tag{13}$$

Note (compare with equation 4) that the corner frequency is $2\pi/(\alpha+\beta)$ and that the variance is $\gamma^2 f(1 - f)$.

The calculations we have made so far have assumed a membrane with only a single channel. We may generalize immediately, however, to the membrane containing N independent and identical channels because the mean conductance is N times the conductance of a membrane with only one channel, and the spectral density (as well as covariance function) is also N times larger.

To use the results developed above, we would measure fluctuations from a membrane containing our channels and then compare the shape of the resulting spectrum (Fig. 3A) with that predicted by equation 13. If the fit were satisfactory and we could accept the theoretical description, then equation 13 would be used to determine quantities such as the single channel conductance and the mean length of time the channel stays open which is, as further calculations would show, $1/(\alpha+\beta)$. In this particular case, the differential equation describing our system is a very simple one, but even in more complicated situations the procedure is always the same. The molecular theory is formalized in terms of differential equations describing the probability of finding the system in any state and these are used to calculate the covariance function which leads to the spectral density. The spectral density can then be used to compare with experimental data.

In order to gain useful information from noise, one must always form a link between molecular events and the spectrum. This link must be some sort of theory. Spectra can therefore only be interpreted in terms of a theory; no theory-free way of extracting information from spectra is possible.

MISCELLANEOUS CONSIDERATIONS

Whenever fluctuation analysis is to be used, the first step is to identify all of the sources of noise contributing to the experimental records. In a membrane system, noise sources generally include equipment contributions (usually a voltage clamp, and sometimes such things as building vibrations) and fluctuations produced by the presence of various channel types and also such sources as damage resulting from microelectrode penetration. It usually is possible to control the situation well enough that a single source of noise, one population of channels being studied, can be varied by experimental manipulations. When the channels are caused to open, if there is not an appreciable noise obvious in the experimental records, no amount of fancy analysis will generally extract any useful information. An important step in every experimental application of fluctuation analysis, then, is to be sure that fluctuations are present and that they arise from only the single source one wishes to investigate. This does not mean that other sources of noise, equipment noise for example, need to be entirely absent, but only that they be smaller than the noise source we wish to investigate. As long as the equipment noise contribution does not change appreciably when the channels are turned on, it is possible to just subtract a spectrum of equipment noise only from the equipment plus channel produced fluctuations in order to get the spectrum of conductance fluctuations: spectra from independent sources simply add.

An important source of noise in many applications is that which arises from the thermal agitation of charge carriers in resistance. This type of noise is termed thermal or Johnson-Nyquist noise and the thermal noise current has a spectral density, for a pure resistor R, that is given by the equation

$$W(f) = 4kT/R \qquad\qquad\qquad (14)$$

where k is Boltzman constant and T is the absolute temperature. Because all frequency components are equally represented, this sort of noise is called "white". For room temperature kT has the value 4.06×10^{-21} watts seconds. A one megohm resistor then produces a thermal noise current with a white (i.e., flat) spectrum and a density of 16.24×10^{-27} amps^2 sec. Each frequency band contributes to this current noise, so the total noise over a 1 KHz band width is 16.2×10^{-24} amps^2. This is an RMS current noise of 4×10^{-12} amps or 4 pA.

The most efficient way to calculate spectra is with a particular Fourier decomposition algorithm, known at the FFT (fast Fourier transform), that makes efficient use of the inherent binary representation of numbers in computers. A small laboratory computer can calculate the spectrum for 256 data points in less

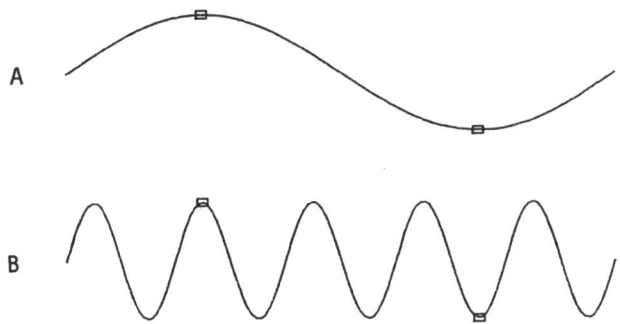

Fig. 4. Illustration of aliasing.

A. Represents one cycle of highest frequency component in a noise sample permitted by sampling theorem. This component has a frequency of 500 Hz in the example given in the text. Two samples, which happened to fail at the peak and trough of the sine wave, are indicated by the squares. In the next example, that sample rate was 1000/sec, or 2 samples per cycle of a 500 Hz sine wave.
B. Represents a 2.5 Hz sine wave that was sampled twice during 2 msec long sample illustrated here. Note that these samples happened to occur at a peak and a trough; thus, were this 2.5 KHz component present in noise sample, it would, in this example, be indistinguishable from the 0.5 KHz component. Antialiasing filters remove frequency components about that permitted by the sampling theorem (here 500 Hz) to eliminate this "spilling over" from high frequency components into the spectrum being calculated.

than a quarter of a second so that calculation of spectral densities is, on the human time scale, instantaneous. Covariance functions are usually less efficient to calculate directly (by equation 5) because they require a large number of relatively slow

multiplication steps. Usually the most rapid way of calculating a
covariance function is by using an FFT first to calculate the
spectrum and then to Fourier transform this spectrum to obtain the
covariance function. FFT programs are available for all modern
laboratory computers.

Because digital computers are most frequently used for
fluctuation analysis, noise records are represented as discrete
sample points. This has an important consequence related to what
is known as the sampling theorem. Suppose that we took a 1 second
long record and sampled once per ms so that we had 1,000 data
points in computer memory. We now start to calculate coefficients
in the Fourier expansion with the FFT or by the discrete analog of
equation 2. The first coefficient calculated would be A_1 and B_1,
the amplitudes of the 1 Hz sine and cosine waves. We could then
calculate A_2 and B_2, the coefficients for the 2 Hz sinusoids. We
continue calculating coefficients until we reach A_{500} and B_{500}.
At this point, we will have used 1,000 pieces of information (our
1,000 sample points) to calculate 1,000 coefficients $A_1 \ldots A_{500}$,
$B_1 \ldots B_{500}$, and we will have exhausted all of the information
contained in our sample. This means that the highest frequency
component about which we can extract information from a signal
samples at a frequency of r samples per second is a frequency $r/2$;
this is a statement of the sampling theorem. For our example, we
would take two samples/cycle for the highest permitted frequency
component (500 Hz) as illustrated in Fig. 4A.

Suppose that the noise in this example also happened to have
an appreciable component with a 2.5 KHz frequency. If we then
sampled as indicated by the squares in Fig. 4B, the 2.5 KHz sine
wave component would be indistinguishable from the 0.5 KHz
component (Fig. 4A). This confusion of components whose frequency
exceeds the limit set by the sampling theorem with lower
frequencies is known as aliasing. In order to eliminate aliasing

and be sure that a spectrum is an accurate representation of the molecular mechanisms, we must use an analog filter to remove those components of the noise above the sampling theorem limit that are present. Such filters are called "antialiasing filters". Because a filter with a Butterworth design is the closest approximation to a square window that does not magnify high frequencies, these filters are generally used in spectral analysis.

I have tried here, with a specific example, to illustrate the steps that one must go through in using fluctuation analysis to learn about molecular mechanisms. One point must be especially stressed: no interpretation of experimental spectra is possible without reference to some specific theoretical structure, either explicit or implicit. There is, therefore, no automatic way of interpreting spectra and the procedure we must use is the general one in quantitative analysis which requires the testing of data against the predictions of specific postulated mechanisms. When care is taken, however, fluctuation analysis is an important alternative technique to gain information about channel systems as the succeeding papers will illustrate.

REFERENCES

Many authors have calculated the spectrum (equation 4) for the system indicated in scheme (7), or its equivalent. The first probably was:

Kenrick, G. W., 1929, The analysis of irregular motions with
        applications to the energy frequency spectrum of static and
        of telegraph signals, Phil. Mag., Series 7, Volume 7:176.

The use of Fourier transforms and the characterization of noise by spectral analysis is clearly and succinctly treated in:

Aseltine, J. A., 1958, "Transform Method in Linear System
     Analysis," McGraw Hill, New York, pp. 1-293.

The derivation of the differential equations such as equation 8
(Markov jump processes) are clearly treated in:

Bailey, N. T. J., 1964, "The Elements of Stochastic Processes,"
     Wiley & Sons, New York, pp. 1-245.
Gnedenko, B. V., 1962,"The Theory of Probability," Chelsea
     Publishing Company, New York, pp. 1-523.

A unified and systematic treatment of multistate generalizations
of scheme 7 is presented in

Neher, E., and Stevens, C. F., 1977, Conductance fluctuations and
     ionic pores in membranes, Ann. Rev. Biophys. Bioeng. 6:345.

The standard source for the description of FFT and data processing
is:

Bendat, J. S., and Piersol, A. G., 1971, "Random Data: Analysis
     and Measurement Procedures," Wiley-Interscience, New York,
     pp. 1-397.

NONSTATIONARY NOISE ANALYSIS OF MEMBRANE CURRENTS

F.J. Sigworth

Biophysic Chemistry
Max-Planck Institute
D-3400 Göttingen, Federal Republic of Germany

INTRODUCTION

The simplest way to study membrane current fluctuations from ionic channels would be to place the channels into a steady, partly activated state and to monitor the fluctuations in the current around its steady mean value. In practice, an experiment of this kind is often not possible because of properties of the channels themselves. Acetylcholine-receptor channels, for example, show a desensitization process in which the agonist-induced membrane current declines with time. Some voltage-activated channels show a similar kind of behavior: during a steady depolarization, Na^+ channels spontaneously inactivate. These changes in membrane current with time pose both practical and theoretical problems for noise analysis. Since the size of the fluctuations is typically only a few percent of the mean current, a serious practical problem arises whenever a small baseline drift occurs, since drift can seriously contaminate a power spectrum computed from the fluctuations (Fig. 1). In many experimental situations the baseline drift can be estimated and removed from the records; however a theoretical problem still

21

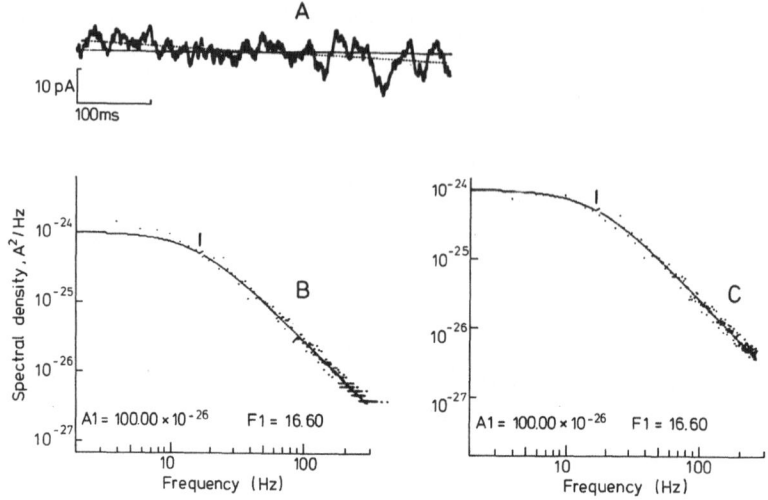

Fig. 1.   Contamination of spectrum by a small baseline drift.
          A. A simulated single-time-constant noise trace, with an
          RMS fluctuation of 4 pA, to which a baseline drift of 10
          pA/sec has been added.
          B. The spectrum computed from 50 such traces. Note that
          the drift causes the low-frequency points in the spectrum
          to lie considerably above the theoretical values.
          C. The spectrum computed from 50 traces, after a
          correction procedure in which a straight line was fitted
          to, and subtracted from, each trace. The two
          lowest-frequency points now lie slightly below the curve,
          due to the effects of the fitting.

remains: how should the fluctuations be interpreted when the mean
current itself changes with time? This chapter presents some
results concerning each of these problems.

For simplicity the usual membrane noise assumptions are made
in this chapter: namely, that the current fluctuations of interest
come from the stochastic opening and closing of
independently-acting, identical ionic channels. The current
fluctuations therefore are assumed to arise from fluctuations in
the number of channels that are open under given conditions.
Three methods of analysis will be described. The first is the
estimation of the time-course of the variance from a set of
current records. From the relationship between the variance and
the mean current the size of the single-channel current, and
sometimes also the total number of channels, can be estimated.
The basic idea in this case is that, given large unit currents,
the fluctuations in the total current are also relatively large.
An extension of the estimation of the variance to also consider
the temporal structure of the fluctuations leads to the second
technique, the estimation of the nonstationary covariance
function. Here the correlation time of the fluctuations is seen
to be related to the time a channel remains open. Finally, a
special case of the theory for the covariance is shown to yield
the power spectrum, which can sometimes be meaningfully
interpreted even when it is computed from nonstationary
fluctuations.

THE VARIANCE OF FLUCTUATIONS

The basic problem in nonstationary fluctuation analysis is to
distinguish the fluctuations from the variations in the mean
number of channels open. In the stationary case this is easy,
since the average number of open channels is constant with time.
One needs only to take the time average of a sufficiently long
trace to obtain a good estimate of the mean, which can then be
subtracted from the trace to yield a record containing "pure"
fluctuations.

In the nonstationary case, in which the mean itself varies
with time, the best approach is to repeat a basic measurement many
times and to use the resulting ensemble of values to estimate the
mean and variance. The basic experimental procedure is to apply a
series of identical, channel-activating stimuli (depolarizations
in the case of voltage-activated channels, for example). At a
fixed time after each stimulus the response (in this case membrane
current) is recorded. Figure 2 shows fluctuations in Na^+ currents
recorded in this way. Each dot in part A of the figure represents
a measurement of the current 1.6 ms after the start of a
depolarization. (Depolarizations were given at 400 ms intervals;
the response to an individual depolarization is shown in part B,
with the point at 1.6 ms marked with a cursor.) Around a mean
value of about 1.3 nA a considerable scatter is seen, with a
standard deviation of about 2%. This is superimposed on a slow
decrease in the mean value, represented here by the straight line
fitted to the points.

A good way to estimate the variance of the fluctuations is to
estimate the squared deviations within groups of a few responses,
and then to average these estimates; if the groups are chosen to
be small enough, the long-term decrease will have a negligible
effect (Sigworth, 1981). The estimate of the variance σ^2 for a
group of n values y_i of the response, is given by the standard
formula

$$\sigma^2 = \frac{1}{n-1} \sum_{i=1}^{n} (y_i - \mu)^2 \tag{1}$$

where μ is the average of the n values. This formula holds only
if the fluctuations among the y_i are uncorrelated; the channels
giving rise to the fluctuations are thus assumed to be completely
"reset" at the time of each stimulus. An important part of the

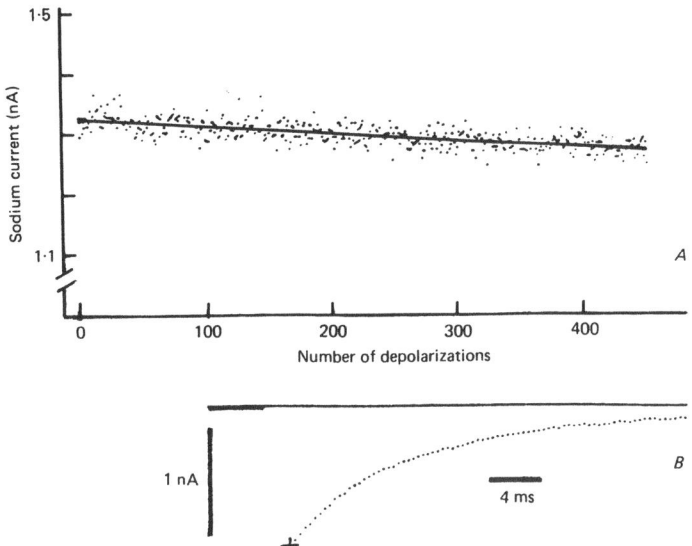

Fig. 2.   Example of fluctuations in the response to a repeated
stimulus. Sodium currents were elicited in a frog node
of Ranvier by depolarizations at 400 ms intervals to -15
mV from a prepulse potential of -65 mV; leakage and
capacitive currents have been subtracted. The magnitude
of the current 1.6 ms after the start of each pulse is
shown in A; a representative current trace is shown in B.
The responses show a random fluctuation of about 2%
around a slow decrease of about 0.01% per pulse (straight
line). (From Sigworth, 1980a).

design of such an experiment is therefore the choice of a
sufficiently long inter-stimulus interval.

Variance estimates as in (1) can be averaged over many groups
to give an estimate of variance with less scatter. Also, the
variance can be estimated at each of many different times after
the stimulus. That is to say, the variance can be computed

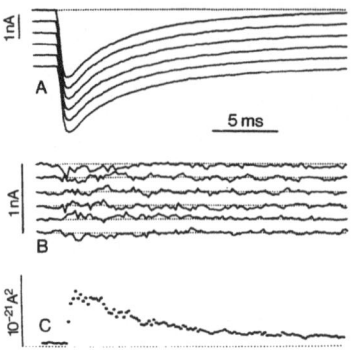

Fig. 3.   Residual fluctuations and variance from a group of
          records of sodium current in the node of Ranvier.
          A.   A setof six records aligned according to the start of
          the depolarizing pulse.
          B.   The residual fluctuations after subtraction of the
          mean of 12 records.
          C.   The variance at each time point, computed from 65
          groups of 6 records.   The currents were elicited by
          potential steps to -5 mV from a prepulse potential of
          -105 mV.   (From Sigworth, 1980a).

point-by-point from a large number of records.   The procedure for
such a computation is illustrated in Fig. 3.   From a set of
successively-evoked current records (A), shown displaced
vertically for clarity, a mean time-course is computed and
subtracted from each record, leaving residual fluctuations (B).
The scaled sum of the squares of these fluctuation records, when
averaged with sums from other groups, gives an estimate of the
time-course of the variance (C).

     In practice the variance includes contributions from two main
sources.   One is the "irrelevant" background noise which arises
mainly from the instrumentation, either directly or indirectly

through its effect on voltage-dependent channels. It also arises
in part from thermal noise in the membrane conductances. The
other, desired, component of the fluctuations arises from channel
activity. The background noise is usually presumed to be
uncorrelated with the channel noise. If so, then its variance can
be estimated in the absence of channel activity, and can be
subtracted from the total. It should be noted, however, that in
many voltage-clamp situations the magnitude of the background
noise changes appreciably with the membrane conductance induced by
channels, complicating the subtraction procedure. The conductance
dependence arises mainly because a voltage-clamp system acts to
bring the membrane to a noisy potential. The noisy currents that
flow as a result depend on the instantaneous membrane conductance
as well as the voltage dependence of gating of membrane currents.
The most effective solution to this problem is to reduce the
background noise in the voltage clamp to make the variance from
this source small. Another approach is to carefully characterize
the noise sources and subtract the predicted background component
from the variance.

The variance contribution from background noise sources can
also be reduced by appropriate low-pass filtering of the signal
before analysis. The filter cut-off frequency should be chosen
high enough to include essentially all of the spectral components
that are expected from channel gating; however, a high cutoff
frequency results in a large background noise variance. The sharp
cutoff filters normally used for stationary noise analysis
(Butterworth, Tschebycheff, etc.) are not necessarily the best for
nonstationary analysis, because they distort the time course of
rapid events. For obtaining the time course of the variance, a
Bessel filter is better.

The time-dependent variance from N identical channels is given by the expression,

$$\sigma^2(t) = Ni^2 p(t)[1 - p(t)] \tag{2}$$

where σ^2 and the probability of a channel being open p are written as functions of t to emphasize that the equation holds even when these quantities are time-varying. Making use of the fact that the mean current I(t) is given by

$$I(t) = Nip(t)$$

equation (2) can be rewritten in terms of I, rather than p, as

$$\sigma^2(t) = iI(t) - \frac{1}{N} I^2(t) \tag{3}$$

which relates the two experimentally-derived functions σ^2 and I through two parameters i and N. By curve fitting, therefore, the single-channel current and number of channels can be estimated, on the basis of the assumptions mentioned earlier.

Equation (3) describes a parabola with an initial slope equal to i and an intercept at I = Ni and σ^2 = 0. The shape of the curve can be qualitatively understood as follows: Remembering that the variance arises from variations in the number of open channels, no variance is expected when no channels are open (I = 0) or when all of the channels are open every time (I = Ni); the maximum variance is expected when half of the channels, on the average, are open.

Figure 4B shows experimental variance values plotted against the mean current and fitted according to Eq. (3). In this case a maximum of only about half of the channels were open at once, so

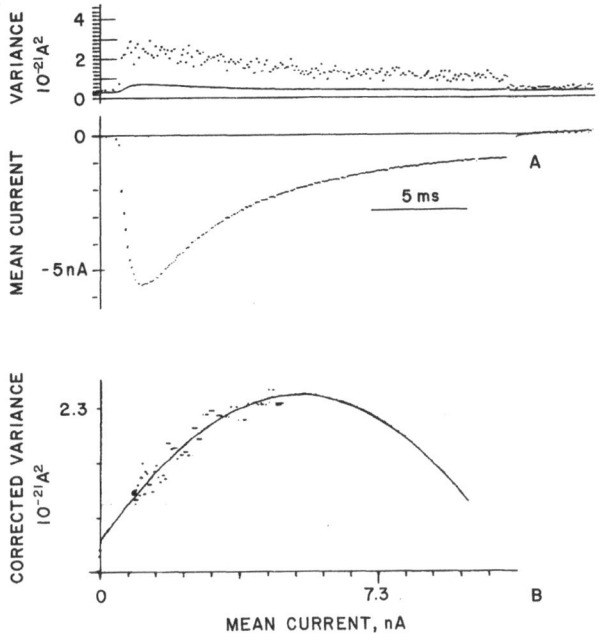

Fig. 4. Variance-mean relationship.

A. Variance and mean computed from 32 groups of four sodium current records, elicited by depolarizations to -15 mV after prepulses to -105 mV. The solid curve is the calculated variance due to background noise sources. B. Plot of the variance versus the magnitude of the mean current; the points have been fitted by a parabola (Eq. 3) with the parameters i = -0.55 pA and n= 20,400. The fit implies that nearly half of the channels are open at the peak of the current.

that the experimental points form only half of the parabola. In many situations the peak fraction of channels open is substantially smaller than this. In these cases reliable values for i can be estimated, but N, which is determined from the

curvature of the plots, cannot be estimated because little
curvature is present.

The analysis of the time-dependent variance was used by
Schwartz (1975) in studying voltage fluctuations in photoreceptor
responses. For voltage-clamp currents the analysis has been used
for sodium currents in axons (Sigworth, 1980a; Neumcke and
Stämpfli, 1982, 1983), sodium currents and inward rectifier
currents in tunicate eggs (Ohmori, 1981), and sodium and calcium
currents in chromaffin cells (Fenwick et al., 1983) and a
pituitary cell line (Hagiwara and Ohmori, 1982).

Situations Where the Simple Analysis is not Valid

The interpretation of the variance becomes more complicated
in situations where the assumptions underlying Eqs. (2) and (3) do
not hold. For example, it was assumed that there was only one
source of variance due to variation in the number of channels that
open when presented with identical stimuli. One process that is
more complicated than this is the one underlying evoked or
spontaneous end-plate currents. The idea would be to compare
miniature epcs to estimate the acetylcholine (ACh) receptor
channel conductance. However, since transmitter quanta are
released from various release sites onto different parts of the
postsynaptic membrane, the observed fluctuations would reflect not
only variations in the number of channels open, but also
variations in the total number of channels near the release sites
and variations in the size of the quanta.

The analysis also becomes more complicated if all the
channels are not identical. A population of channels having a
distribution of single-channel currents but identical time courses
of the opening probability $p(t)$, for example, would give rise to a
variance-mean relationship of the form of (3), but with i being a

weighted average of the various current values (Sigworth, 1980b).
If the kinetics are different, a simple relationship of the form
of (3) for the overall variance does not apply. Instead, the
expected time course of the variance and of the mean must be
evaluated for each population, and the sums of these functions
fitted to the data. A test of this kind for two populations of
sodium channels is described in Sigworth (1981).

A third situation in which the analysis does not hold is the
case of interacting channels. An interaction is defined here as
the influence of one channel on the state of another. If, for
example, the opening of one channel increases the probability of
the opening of other channels, fluctuations will be augmented,
causing the variance, and therefore the apparent single-channel
current to be larger than expected. In general the interaction
between channels would depend on their past history as well as
their present states. This means that the time course of the
variance will also be affected, so that the variance-mean
relationship will not be of the form of Eq. (3). A common
artificial source of interactions is a residual series resistance
in the voltage-clamp system. For membrane conductances that have
a negative-resistance characteristic, such as Na channels, a
series resistance error results in a positive interaction that
increases as the membrane current increases. Relatively large
errors can cause the variance-mean relationship to curve upward,
rather than downward as in Fig. 4.

One way to test for interactions among channels is to block a
fraction of the channels using an irreversible or
slowly-reversible channel blocker. Since blockage lowers the
number (i.e., density) of channels that carry current, the
interactions among these channels would be expected to be smaller.
In the absence of interactions the variance-mean relationship
would have the same form and the same value of i, but different

values of N reflecting the degree of block; with interactions
other changes would be expected. A test of this kind on Na
channels at the node of Ranvier (Neumcke and Stämpfli, 1983) shows
that such interactions may be present.

If interactions are time-dependent, that is if the state of
one channel at a given time influences the behavior of other
channels at later times, then a test for interactions can be made
by using different stimulus protocols to place the channels in
different initial conditions. In the absence of interactions, the
variance-mean relationship should not depend on initial
conditions. A test of this sort, in which the size of a
hyperpolarizing prepulse was varied, was used to test for
interactions among potassium channels in the node of Ranvier
(Sigworth, 1979). The result was consistent with there being no
interactions among the channels.

THE NONSTATIONARY COVARIANCE

The covariance can be looked at as a generalization of the
variance that characterizes not only the size of the fluctuations,
but also their temporal structure. The covariance $C(t_1, t_2)$ is
defined as the average product of fluctuations at the times t_1 and
t_2, the average being taken over an ensemble of experimental
records.

$$C(t_1, t_2) = < x(t_1)x(t_2) > \qquad\qquad (4)$$

where $x(t)$ is the residual fluctuation (like the traces in Fig.
2B) when the mean time-course is subtracted from an individual
record.

The covariance shows how well fluctuations are correlated at different times. Although x(t) has an average value of zero, the product in (4) tends to have a nonzero positive value when t_1 and t_2 are chosen close together. (When t_1 and t_2 are equal, the covariance is just equal to the variance). When t_1 and t_2 are chosen far apart, the product tends to zero as $x(t_1)$ and $x(t_2)$ show less correlation. The time difference that is required for C to drop to near zero reflects the length of time required for a fluctuation in open channel number (for example, a momentary excess of open channels) to disappear or be replaced by an uncorrelated fluctuation. This time, in turn, is related to the rates of opening and closing of channels; thus information about the kinetics of channels can be gained from the covariance.

The practical computation of the nonstationary covariance can be done in a group-wise manner to avoid errors from long-term drift. Analogous to the calculation of the variance in Eq. (1), an unbiased estimate of the covariance from a group of n records is given by

$$C_n(t_1,t_2) = \frac{1}{n-1} \sum_{i=1}^{n} [y_i(t_1) - \mu(t_1)][y_i(t_2) - \mu(t_2)] \quad (5)$$

where the y_1 are current values from the ith record, and the μ are the means of the ny_i values. Estimates from short groups can be averaged to reduce the scatter in the values of C. In practice a large total number of records is required to give a usable covariance function: 100 groups of 4 records were used to compute the function plotted in Fig. 5. The computation of C can take appreciable computer time, since for M time points and N records, NM^2 multiply-add operations are required to produce M^2 values of C.

In the case that the channels are identical, independent and are either open or closed, the kinetic information from the covariance can be expressed in the conditional probability function $p_{11}(t_2|t_1)$. This function is the probability, given that a particular channel is open at a reference time t_1, that the same channel will be open at a different time t_2. The function is useful for kinetic analysis because it describes the behavior of only those channels that are open at the given time t_1; it provides the same kinetic information as would be obtained if, somehow, every channel were opened at t_1 and the subsequent time-course of relaxation were then monitored. Such information from the covariance computed from stationary data has been useful in the past, for example in the case of ACh-induced currents at the motor end plate: although the instantaneous application of ACh (to synchronously force channels into the open state) has not been experimentally feasable, analysis of the covariance (or, equivalently, the power spectrum) has shown the time-course of relaxation from the open state of the channel. The nonstationary covariance similarly shows the time-course of relaxation from being open, but has the added property of characterizing the relaxations starting at a certain reference time. The utility of this property will be demonstrated below.

The conditional probability function is related to the covariance by the equation (see Sigworth, 1981, for a derivation).

$$C(t_1,t_2) = Ni^2[p(t_1)p_{11}(t_2|t_1) - p(t_1)p(t_2)] \qquad (6)$$

where the parameters N, i and p(t) have the same meanings as before. Since $p(t) = I(t)/Ni$, we can eliminate p and rewrite (6) as

$$C(t_1,t_2) = iI(t_1)p_{11}(t_2|t_1) - \frac{1}{N} I(t_1)I(t_2) \qquad (7)$$

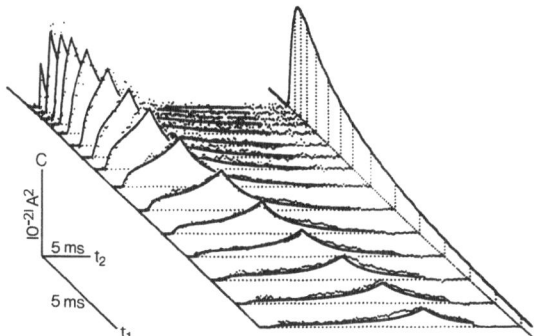

Fig. 5. Plot of the covariance function $C(T_1,t_2)$ computed from
390 Na current records elicited by depolarizations to -5
mV in a frog node of Ranvier. The continuous curves are
the prediction of a specific kinetic scheme, while the
points are experimental values. To the right is plotted
the time course of the Na conductance, shown as a
function of t_1. (From Sigworth, 1981a).

which reduces to Eq. (3) when $t_1 = t_2 = t$. From the variance-mean
analysis of a set of fluctuation data, i and N can be estimated;
using these values and a record of the mean-current time course
I(t), P_{11} can be computed from the experimental values of C,

$$P_{11}(t_2|t_1) = \frac{C(t_1,t_2)}{iI(t_1)} - \frac{I(t_2)}{Ni} \tag{8}$$

as the covariance, normalized depending on the mean current at t_1,
with an added "correction term" equal to the probability of a
channel being open at t_2.

Properties of p_{11}

The properties of P_{11} reflect the kinetic steps involved in opening and closing the channel, i.e. the gating process. Suppose that a channel behaves in such a way that, given that it is open at time t_1, its behavior after t_1 does not depend on past history, but depends only on the time displacement. In other words, p_{11} can be written as

$$p_{11}(t + \tau) = \phi(\tau) \tag{9}$$

where $\phi(\tau)$ is a function that depends only on the time displacement τ, which is here assumed to be positive. This properly requires that the gating process be a Markoff process, in which the future behavior depends only on the particular state the channel is in at any given time, regardless of how it was placed into that state. Further, the property (9) requires that there be only one state corresponding to the "open channel" conformation. Channels with two or more states corresponding to the open channel do not generally show this property, since the decay from being open depends on which "way" the channel was open in the first place. If the relative probability of being in the various open states changes with time, then $p_{11}(t_1 + \tau | t_1)$ will depend not only on the displacement τ, but the reference time t_1 as well.

The test for a single open state is important for the understanding of channel function. The existence of more than one open state implies that the channel molecule can undergo significant internal changes without affecting the ion flow. In principle this sort of behavior can be distinguished from the similar results arising from multiple populations of channels having different kinetics. Lumped together, these channels would not satisfy (9) even if individual populations did. The variance-mean relationship from such a mixture of channel types would differ from the simple relationship (3) as discussed above, though these differences might be hard to detect.

Fig. 6 shows a test for the property (9) in node of Ranvier sodium channels. The dashed curves are the prediction of a gating scheme with one open state; they show the same decay from being open at each t_1 value. The data points, computed according to eq. (8), however show a slower relaxation at later t_1 values. Similar behavior is shown by the solid curve which is the prediction of a scheme with two open states. The second open state has a longer lifetime and is populated preferentially at later times. Thus the relaxation is slower at later values of t_1. This scheme is only one of many possible schemes having multiple open states (or multiple, heterogeneous populations of channels) that could provide an acceptable fit to the data. It can therefore only be concluded that one-open-state theories cannot work, since they all predict the property (9).

Limitations on This Analysis Technique

The theory for the nonstationary covariance is based on the same assumptions that were made for the variance analysis, and is therefore subject to errors from systematic drifts from inhomogeneous populations of channels, or if channels interact.

POWER SPECTRA FROM NONSTATIONARY DATA

The reader might well be asking himself at this point whether it would be possible to compute a conventional power spectrum to obtain kinetic information from the fluctuations, rather than dealing with the complexities of the nonstationary covariance. In fact a power spectrum can be computed from records after

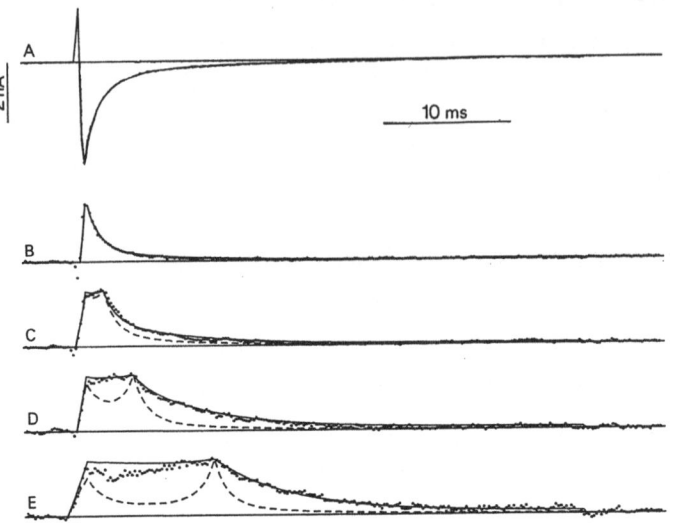

Fig. 6. Conditional-probability functions computed from the
         nonstationary covariance.

         A. Mean current time course. After a prepulse to -105
         mV, a brief (0.4 ms) activating pulse to +125 mV
         produced a brief outward current, which became inward
         when the potential was stepped back to +5 mV. The dashed
         curves demonstrate one way that the inactivation
         time-course can be separated into fast and slow
         components.

         B-E. Conditional probabilities $p_{11}(t_2|t_1)$ computed
         according to eq. (8) for the four different values of t_1
         given. The dashed curves are the prediction of a kinetic
         scheme with one open state; note that its decay is too
         rapid at the larger t_1 values. The solid curves are the
         predictions of a two-open-state scheme. (From Sigworth,
         1981a).

subtraction of the mean (as in Fig. 1). Conti et al. (1980), have
introduced this technique in analyzing Na current fluctuations at
the node of Ranvier. They computed the average power spectrum

from the differences of pairs of records. This caused the mean
component to be removed, but left the fluctuations from both
records in the difference; they therefore divided the resulting
spectral estimates by two to recover the average spectrum of
fluctuations in a single record. This procedure is analogous to
the group-wise estimation of the variance (Eq. 1) and covariance
but with $n = 2$.

This subtraction procedure is very effective with sets of
records having a repeatable, slow baseline drift, as in the sodium
current records of Conti et al. It turns out, however, that not
only slowly drifting records, but also highly nonstationary data
can be analyzed in this way. Figure 7 shows the computation of a
spectrum from Na^+ current records that included most of the time
course of activation and inactivation. The resulting spectrum
shows no sign of contamination from the inactivation process.

The interpretation of spectra obtained by the subtraction
procedure is very similar to that of spectra computed from
stationary fluctuations. Predictions of the power spectrum can be
computed from specific kinetic schemes (Conti et al., 1980) by
making use of the fact that the power spectrum is the Fourier
transform of the one-dimensional covariance, which, in turn, is
just a time average of the nonstationary covariance that we
considered in the last section.

$$C(\tau) = \frac{1}{T} \int_{o}^{T} C(t_1, t + \tau)dt$$

Thus the power spectrum can be predicted by first evaluating
the nonstationary covariance. A simpler procedure can be followed
however, if the underlying channel kinetics show the property of
eq. (9) - that is, if the decay of channels from the open state

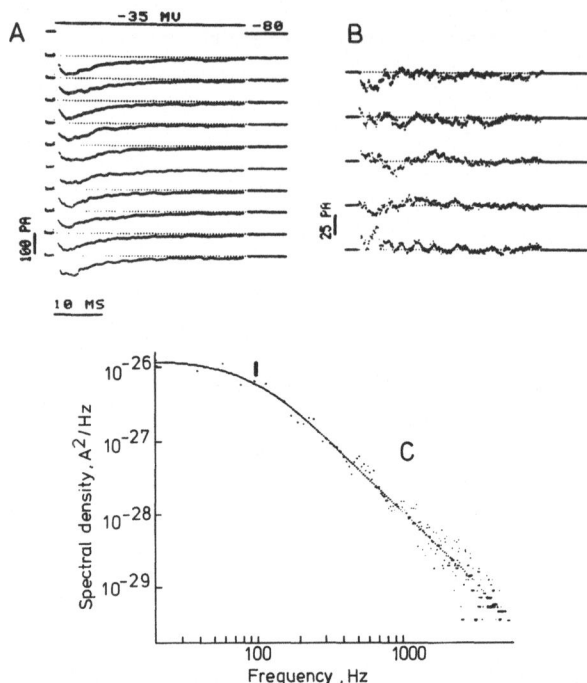

Fig. 7. Sodium current fluctuations in a chromaffin cell during
voltage-clamp steps to -35 mV. A bovine chromaffin cell
of diameter approximately 12 μm was voltage clamped using
a single pipette (Hamill et al., 1981) containing a Cs[+]
internal solution to block K[+] channels; the external
medium contained Co[++] to block Ca[+] currents.

A. Successive current records after leak subtraction.
The voltage trace is shown at the top.

B. Differences between pairs of records, shown with 4
times the vertical scaling.

C. The average spectrum, computed from the differences.
The curve is a Lorentzian with amplitude 1.2 x 10[-25]
A^2/Hz and cutoff frequency 100 Hz.

always shows the same time course. The theory is developed in
Sigworth, 1981b; the resulting expression for the power spectrum
$G(f)$ is

$$G(f) = i\bar{Ip}\Phi(f) - \frac{1}{N} G_I(f) \tag{10}$$

where $\bar{p}$ is the probability of a channel being open, averaged over
the length of a record and $\Phi(f)$ is the Fourier transform of the
decay function $\phi(\tau)$ (see Eq. 7). $G_I(f)$ is the power spectrum of
the time course $I(t)$ of the mean current.

Eq. 10 is very similar to the usual expression for the
spectrum from stationary fluctuations; the only differences are
that the time-averaged probability $\bar{p}$ is used instead of the
constant probability p, and that a second, frequency-dependent
correction term is present. In the stationary case the second
term is also present, but it contributes only at $f = 0$ and is
therefore usually ignored. If I varies during the record, on the
other hand, the correction contributes at non-zero frequencies as
well. The relative size of the second term is small, however, if
the variations are small or if p is small (e.g., $p < 0.1$)
throughout the experimental record. The origin of the second term
can be understood from the fact that when p approaches 1 the
variance of the fluctuations vanish (see Eq. 2). The effect of
the second term on the spectrum is a depression of spectral
values, especially at the lower frequencies.

This second, correction term can be estimated and removed
from an experimental spectrum. This allows the resulting
spectrum, which is proportional to $\Phi(f)$, to be interpreted in the
same way as the spectrum from stationary fluctuations. For
example, if the decay function $\phi(\tau)$ shows a single-exponential

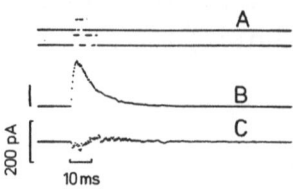

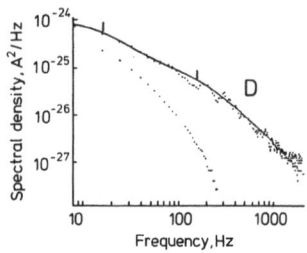

Fig. 8.   Simulated membrane currents and the spectrum computed
          from the differences between pairs of records.
          A. Typical simulated single-channel events.
          B. A "current record" obtained by summing 1000 events
          like those shown in A.
          C. The difference between two records. The average
          spectrum was computed from 50 such differences.
          D. The corrected spectrum, obtained by adding the
          correction term (shown as the dotted curve) to the
          average spectrum. The smooth curve is the predicted
          two-component spectrum. (From Sigworth, 1981b).

relaxation, $\Phi(f)$ will be Lorentzian, as will be the corrected
spectrum. The correction is estimated by taking the power
spectrum of the mean current time course, and then scaling it by
$1/N$. N, in turn, can be estimated by fitting the variance-mean
relationship.

Fig. 8 demonstrates this procedure with simulated data. Single-channel records were simulated by means of random numbers to correspond to the kinetic scheme

$$S_0 \underset{0.55}{\overset{0.37}{\rightleftarrows}} S_1^* \overset{0.18}{\longrightarrow} S_2$$

where rate constants are given in $(ms)^{-1}$. At the beginning of each trace, all channels are in state S_0. The decay function $\phi(\tau)$ is given by the probability of a channel being in the open state S_1 at time τ given that it was in S_1 at time 0. Its form is

$$\phi(\tau) = \frac{1}{2} \exp(-\tau/1 \text{ ms}) + \frac{1}{2} \exp(-\tau/10 \text{ ms})$$

where the two terms reflect the fast relaxation between states S_0 and S_1 and the slower decay into state S_2 respectively. The spectrum function $\Phi(f)$ similarly has two Lorentzian components; it is shown, scaled by the factor Ni^2, as the smooth curve in Fig. 8D. The average spectrum computed from 50 pairs of records was corrected by adding the spectrum of one unsubtracted record scaled by $1/N$ (dotted curve). The correction was appreciable in this case because p reached a relatively large value (0.43) at the peak of the current. The resulting corrected spectrum agrees well with the theoretical curve.

Armed with this theory, let us attempt to interpret the experimental data shown in Fig. 7. The variance-mean relationship (not shown) had no obvious downward curvature, implying that p is small at the peak current, so that the correction is expected not to be important. The good fit with a single Lorentzian function implies that a single relaxation mode predominates. This suggests, in turn, that only one open state is populated, and that the relaxation reflects a channel open time $\tau = 1/2\pi f_c \cong 1.6$ ms. The lack of an obvious slow component corresponding to the inactivation time constant suggests that there are no "groups" of opening events, as in Fig. 6, but rather that a given channel rarely reopens during a trace. These inferences are borne out by single channel recordings that have been made under similar conditions (Fenwick et al., 1982).

Finally, it should be emphasized that this sort of analysis of power spectra is based on the assumption of one open state of the channel (or, more generally, the Markoff process assumption of Eq. 9). When there are multiple open states, the spectrum reflects a superposition of the conditional probabilities of being in the various states, and must be predicted by means of the nonstationary covariance.

The Spectrum from a Single Nonstationary Record

Up to this point we have assumed that a large number of current records are available for analysis, having been elicited by identical, repetitive stimuli. Sometimes such a set of records cannot be obtained, and all that is available is a single record with a time-varying mean current, as in Fig. 9A. Provided that such a record is long enough, an acceptable spectrum can nevertheless be obtained which can be interpreted in terms of the theory that was just presented.

Figure 9 demonstrates this procedure. Part A of the figure shows a simulated current record from drug-activated channels having a mean open time $\alpha^{-1} = 1$ ms and a single channel current of 1 pA, when they are exposed to a time-varying agonist concentration of the form $C_o \exp(-t/\tau_a)$ with $\tau_a = 0.5$ sec. One-to-one binding of agonist to receptor is assumed, and the concentration is assumed to be low, so that p is very small. The current signal is high-pass filtered (Fig. 9B) and all but the first part of the filtered trace is divided into segments. The

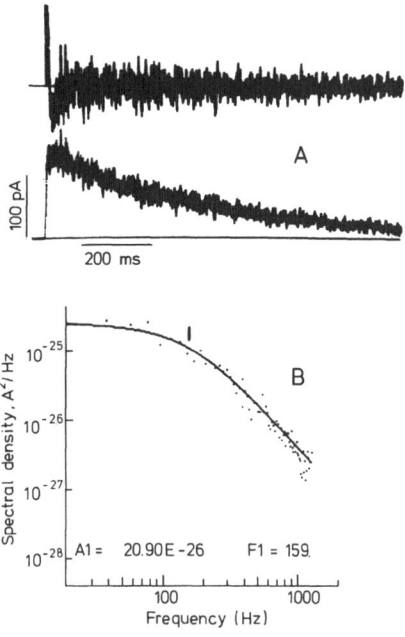

Fig. 9. Power spectrum computed from a single nonstationary

current record.

A. Lower trace, shows the simulated current record, such

as would be elicited by drug-activated channels having a

1 ms open time in the presence of an exponentially

decaying agonist concentration. The upper trace shows

the same record at twice the vertical scale after

high-pass filtering (8 Hz, 2-pole Butterworth response).

B. Power spectrum calculated from the high-pass filtered

record. The record was sampled at 0.2 ms intervals and

divided into ten 512-point blocks. The first block was

discarded because of the filter transients. The average

of the 9 spectra computed from the remainder of the trace

is shown here. The mean current during that part of the

trace was I = 55 pA. Using eq. 10, the predicted

spectrum is shown as the curve. The point at 20 Hz is

low by a factor of 2 because of the high-pass filter

characteristic.

power spectra computed from the segments are averaged, yielding
the result in part C.

The form of the resulting spectrum is predicted by Eq. (10),
where, in this case, the average current I and, if required, the
correction term $G_I(f)$, are evaluated as averages from all portions
of the record that were used for computing the spectrum. Since p
is known to be small, the correction term is ignored, and the
spectrum is expected to have the single-Lorentzian form

$$G(f) = i\bar{I}\Phi(f) = \frac{4i\bar{I}}{\alpha} \frac{1}{1 + (f/2\pi\alpha)^2}$$

This function is shown as the smooth curve in the figure. The
lowest-frequency data point (at 8 Hz) is low by a factor of two
because of the high-pass filtering.

High-pass filtering is important in this case since it
prevents very low frequency components from contaminating the
spectrum by an aliasing process. Such low frequency components
would appear as baseline drifts in short segments and would
contribute a spurious $1/f^2$ component to the spectrum, as in
Fig. 1. The high-pass filter should have a reasonably sharp
rolloff (at least 12 db/octave, i.e., 2 poles), although
high-order filters display long transient responses which imply
long delays before the initial transient dies down. A good
compromise is a 2-pole Butterworth filter, whose transient
response decays with a time constant of $1/2\pi f_c$, where f_c is the
cutoff frequency. A further improvement in settling time can be
obtained by electronically switching the filter characteristic to
rapidly charge the internal capacitors (Conti et al., 1980). This
results in a smaller initial transient, but some care should be
taken because the remaining transient wave has a random amplitude,

depending on the instantaneous signal amplitude at the time of switching the filter.

Under many conditions a high-pass filter will not suffice to separate slow trends in a record from the rapid fluctuations of interest. This occurs when the "trend" in the record contains frequency components that are comparable to those in the frequency range of interest that arise from the fluctuations themselves. If the form of the trend function is known, a least-squares fitting procedure can be used to remove the trend. This is actually a sophisticated kind of filtering, and like a high-pass filter, generally distorts the low-frequency values of the resulting spectrum. The distortion arises because the fluctuations in the record perturb the fit in a way to reduce the remaining fluctuation amplitude; the magnitude of the resulting errors can be estimated by computing the effect of fluctuations at each frequency on the fitting parameters, and then in turn the effect of perturbations in the fitting parameters on the spectral values.

REFERENCES

Conti, F., Neumcke, B., Nonner, W., and Stämpfli, R., 1980, Conductance fluctuations from the inactivation process of sodium channels in myelinated nerve fibres, J. Physiol., 308:277.

Fenwick, E. M., Marty A., and Neher, E., 1982, Sodium and calcium channels in bovine chromaffin cells, J. Physiol., 331:599.

Hagiwara, S., and Ohmori, H., 1982, Studies of calcium channels in rat clonal pituitary cells with patch electrode voltage clamp, J. Physiol., 331:231.

Hamill, O. P., Marty, A., Neher, E., Sakmann, B., and Sigworth, F. J., 1981, Improved patch-clamp techniques for high-resolution current recording from cells and cell-free membrane patches, Pflügers Arch., 391:85.

Neumcke, B., and Stämpfli, R., 1982, Sodium currents and
    sodium-current fluctuations in rat myelinated nerve fibres,
    J. Physiol., 329:163.

Neumcke, B., and Stämpfli, R., 1983, Alteration of the conductance
    of Na^+ channels in the nodal membrane of frog nerve by
    holding potential and tetrodotoxin. Biochim. Biophys. Acta,
    727:177.

Ohmori, H., 1981, Unitary current through sodium channel and
    anomalous rectifier channel estimated from transient current
    in the tunicate egg, J. Physiol, 311:289.

Schwartz, E. A., 1975, Rod-rod interaction in the retina of the
    turtle, J. Physiol., 246:617.

Sigworth, F. J., 1980a, The variance of sodium current
    fluctuations at the node of Ranvier, J. Physiol., 307:97.

Sigworth, F. J., 1980b, The conductance of sodium channels under
    conditions of reduced current at the node of Ranvier, J.
    Physiol., 307:131.

Sigworth, F.J., 1981a, Covariance of nonstationary sodium current
    fluctuations in the node of Ranvier, Biophys. J., 34:111.

Sigworth, F.J., 1981b, Interpreting power spectra from
    nonstationary membrane current fluctuations, Biophys. J.,
    35:289.

ANALYSIS OF MEMBRANE PROPERTIES USING EXTRINSIC NOISE

Richard T. Mathias

Department of Physiology
Rush Medical College
Chicago, IL 60612, USA

INTRODUCTION

Biological membranes have been studied using a wide variety
of techniques. One of the newest and most exciting of these
techniques is to observe the random component of membrane current
which occurs in response to the opening and closing of intrinsic
membrane channels. A closely related approach is to drive the
channels with an extrinsic signal of random time course and
observe the correlation between fluctuations in membrane current
and fluctuations in the input voltage. These approaches are
related in the sense that each requires frequency domain analysis
and each uses the same model of channel gating to interpret the
data. They differ in that each experiment emphasizes a different
aspect of the model for gating; but in situations where both
techniques can be applied, this difference makes them
complementary.

There are many situations where one cannot record intrinsic
fluctuations and must resort to an extrinsic driving signal. If
we are interested in the properties of a tissue, then the

49

intrinsic membrane noise is difficult to resolve owing to the
sheer number of channels. Indeed, tissue morphology often
contributes importantly to natural function, so we should not
ignore the role of structure in electrical events. For example,
although ionic channels are the basis of the cardiac action
potential, we will never completely understand this electrical
event without understanding the role of intercellular clefts in
regulating ion concentrations, and the role of gap junctions in
mediating propagation.

Cellular volume regulation in the crystalline lens is
another example of the importance of structure to a natural
function. Volume is directly regulated by an electrogenic sodium
potassium exchange system located primarily in the membranes of a
single layer of epithelial cells. Yet this pump is not the only
factor in the volume control scheme. The fiber cells comprising
the mass of the lens have little capacity to actively transport
sodium and potassium, hence they must rely on gap junctions to
keep them in communication with the surface epithelial cell
layer. Moreover, the passive movement of ions across the fiber
cell membranes, into the tiny intercellular clefts, sets up a
standing current loop, a standing potential gradient, standing
ionic concentration gradients and standing water flow. We are
interested in cellular ion balance and volume regulation because
a failure of this system may be a contributing factor in
cataractogenesis; but in order to detect the cause of such a
failure, we must understand the role of each structural component
in the overall scheme.

In addition to providing information on the electrical
anatomy of a tissue, extrinsic driving signals can be used to
directly study the properties of channels. This method gives
information that is related to, but slightly different from, the
information derived by fluctuation analysis. The power spectrum

of fluctuations is now being vigorously exploited as a probe of channel properties, yet many of the techniques necessary to record power spectra are common to the measurement of linear frequency domain impedance data. One expects that extrinsic noise analysis will begin to play a larger role in the study of channel kinetics and gating.

The purpose of this paper is to describe the techniques and artifacts associated with linear, frequency domain analysis of a membrane or tissue. The application of these techniques to a structural analysis of a tissue is reviewed in Eisenberg and Mathias (1980). The application of these techniques to the analysis of membrane channels will be discussed in this paper.

Frequency Domain Analysis

The theory of frequency domain analysis is well described in many basic textbooks (Desoer and Kuh, 1969; Bendat and Piersol, 1971; Papoulis, 1977), so this section will present a few practical considerations and a brief definition of terms. For an experimental scientist, the frequency domain analysis of a time domain signal involves only finite amplitude signals of finite duration T. Fourier theory shows that any such signal can be represented as a series of sine waves whose frequencies are multiples of the basic harmonic $\omega_0 = 2\pi/T$, thus any signal we measure may be represented as

$$v(t) = \sum_{n=0}^{\infty} V_n \sin (n\omega_0 t + \theta_n) \qquad 0 \leq t \leq T \qquad (1)$$

It is worth noting that the right hand side of equation (1) only describes v(t) on the interval $0 \leq t \leq T$. If the signal v(t) exists at times outside of the interval during which we have

recorded it, the right hand side of (1) will incorrectly
represent it at these times. The right hand side of (1) in fact
gives a repeating or periodic representation of the signal for
t < 0 or t > T, even if the physical signal is not periodic.

In any experiment we can only record and analyze a finite
number of data points. If $v(t)$ has been digitized, then we will
have N data points which determine N/2 of the amplitudes V_n, and
N/2 of the phase angles θ_n. These amplitude and phase points,
evaluated at N/2 frequencies, uniquely determine the N time
points we have recorded, however, if there are high frequency
oscillations in the signal $v(t)$, which occur between time
samples, then we will not be able to resolve these frequencies
without increasing the sample rate. In general, the lowest
frequency that can be resolved is $\omega_o = 2\pi/T$ and the highest
frequency or bandwidth will be $N\omega_o/2$, where the sample interval
$\Delta t = T/N$.

If we apply to a linear circuit a periodic input current, of
arbitrary time course within the period T, then the following
relationships for the input current $i(t)$ and voltage response
$v(t)$ are of basic importance.

$$i(t) = \sum_{n=0}^{\infty} I_n \sin (n\omega_o t + \Psi_n) \tag{2}$$

$$v(t) = \sum_{n=0}^{\infty} |Z(jn\omega_o)| I_n \sin (n\omega_o t + \theta_n) \tag{3}$$

Where,

$$|Z(jn\omega_o)| = V_n/I_n \tag{4}$$

$$\phi_z(n\omega_o) = \theta_n - \Psi_n \tag{5}$$

and $Z(j\omega)$ is the complex impedance function for the circuit,
determined theoretically by taking the Fourier transform of the
differential equation for the current voltage relationship.
$|Z(jn\omega_o)|$ is the amplitude of $Z(j\omega)$ and $\phi_z(n\omega_o)$ is the phase
angle of $Z(j\omega)$, defined by the arctangent of the ratio of the
imaginary part of $Z(j\omega)$ divided by the real part of $Z(j\omega)$. From
a practical standpoint, equations (4) and (5) can be viewed as
defining the impedance function $Z(j\omega)$: the impedance of a circuit
is its response to sinusoidal inputs. The magnitude of $Z(j\omega)$ is
simply the magnitude of the voltage response for sinusoidal input
current of frequency ω and unity amplitude; the phase
angle of $Z(j\omega)$ is the normalized time delay or phase delay
between a peak in the output sinusoid and a peak in the input
sinusoid.

There were two constraints placed upon the relations given
in equations (2) through (5). First, the circuit must be
approximately linear (see the section "Linear Channel Properties"
for a discussion of nonlinear effects) and second, $i(t)$ must be
periodic of period T. In many experiments the input is not
periodic, and T simply represents the time over which we have
observed the input and output signals. In this situation, at
times shortly after $t = 0$, $v(t)$ will in part be responding to the
input current that was applied at times prior to $t = 0$.
Moreover, at times after $t = T$, $v(t)$ will continue to respond to
the input current applied prior to $t = T$. These "end effects"
will introduce errors into the relations given in equations (2)
through (5); however, in the section "Computation of a Transfer
Function" a method of averaging Fourier transforms of the current
and voltage is described, and this averaging will remove "end
effects".

Extrinsic Inputs

A variety of extrinsic inputs have been used to study
membrane channels. Classically, large voltage steps have been
used (Hodgkin and Huxley, 1952) to study the nonlinear response
of channels. And small amplitude sine waves (Cole, 1972) have
been applied to measure the linear membrane admittance. More
recently, investigators have used extrinsic noise to study the
linear properties of tissues and membranes (Fishman et al., 1981;
Mathias et al., 1979; Fernandez et al., 1982).

Figure 1 illustrates most of the extrinsic signals that have
been used to drive biological preparations. The left hand panels
illustrate the time domain input. Each time domain signal is
composed of a series of sine waves of varying amplitude and phase
angle. The right hand side of Figure 1 illustrates the amplitude
spectrum for the sinusoidal components of the time domain input.
Because the most sensitive technique for studying linear systems
is frequency domain analysis, the utility of each input should be
evaluated in terms of its frequency domain spectrum.

Probably, the most extensively used input signal is the step
or square wave illustrated in panel A of Figure 1. The advantage
of a step input is that many sine waves are simultaneously
applied to the system, but the disadvantage is that their
amplitude spectrum falls off as one over the frequency. Thus,
step inputs heavily weight low frequency behavior. If the system
has interesting behavior at high frequencies, then a step is a
poor choice for an input since the system will not be
significantly stimulated at high frequencies.

One way of ensuring the system is significantly stimulated
at all frequencies of interest is to apply a sine wave at each
frequency. Each sine wave represents one point in the frequency

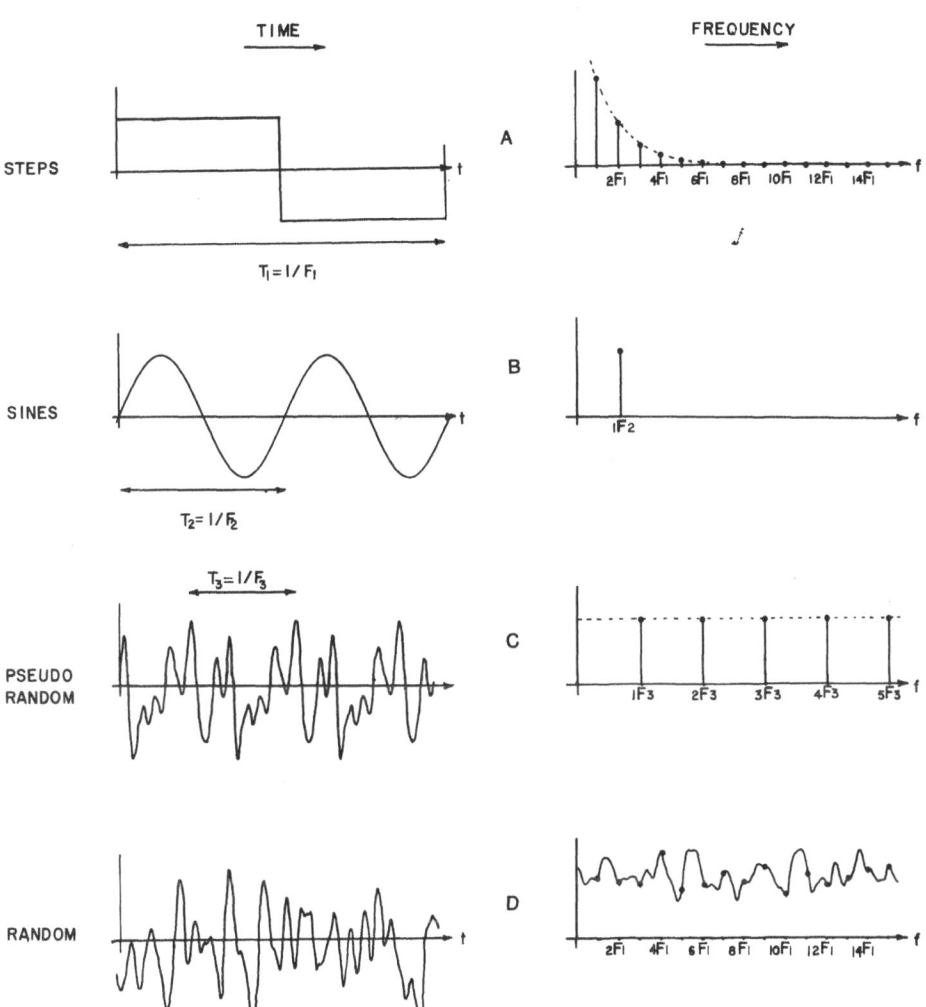

Fig. 1.   Extrinisic signals commonly used as an input waveform in
electrophysiological experiments.   The left hand panel
shows the time domain signal and the right hand panel
shows the amplitude spectrum of the sinusoidal
components of the time signal (i.e., the magnitude of
its Fourier transform).

domain, hence one must perform as many experiments as frequencies
he wishes to study. This is a severe disadvantage when studying
biological preparations, since living tissue usually begins to
deteriorate as soon as it is excised from an organism.

Panel C of Figure 1 illustrates a more modern choice of
input function called a pseudo random signal. One can clearly
see that this is not a random signal but rather it is periodic.
Within one period, however, the signal appears random, hence the
name pseudo random. This signal is composed of equal amplitude
sine waves, in random phase, spaced in frequency at one over the
period of the time input. By increasing the period, we can
achieve an arbitrarily dense frequency spectrum which has equal
energy at each frequency of stimulation. And by increasing the
number of points we can achieve an arbitrarily large bandwidth of
stimulation. Nevertheless, this signal still has a discrete
frequency spectrum, so we must synchronize our recording to
correspond with the discrete frequencies at which we are
stimulating the system. This turns out to be more difficult than
one might expect, because commercial pseudo random signal
generators put out 2^{n-1} frequency points, whereas commercial
spectral analyzers compute 2^n frequency points. There are some
clever techniques which allow synchronization of input with
recording frequencies (Poussart and Ganguly, 1977; Clausen and
Fernandez, 1981). Nonetheless, this is a technical difficulty
one must solve before using a pseudo random input.

At the bottom of Figure 1 (Panel D), we see a fourth choice
of input signals: a random input called extrinsic noise. Such a
signal has a continuous frequency spectrum which, on the average,
will have equal energy at each frequency within the specified
bandwidth. Thus, whatever frequency components are recorded, we
are sure to be stimulating at those frequencies. However, if we
observe such a signal for a finite period of time, then the

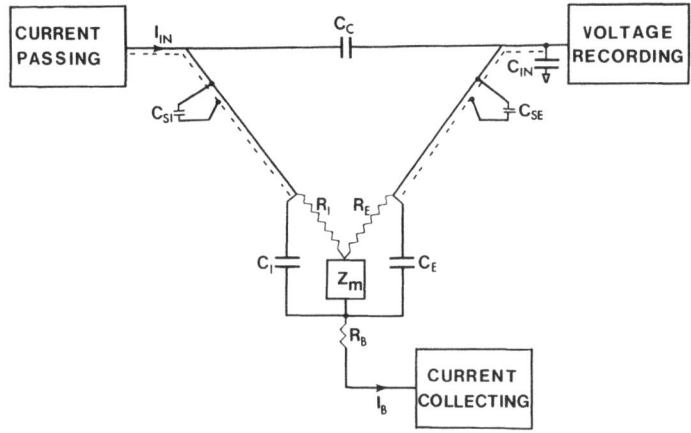

Fig. 2. The equivalent circuit of a two microelectrode

experiment. R_I and R_E are resistances due to the top

of the current passing and voltage recording electrodes

respectively. Z_m is the input impedance of the cell or

tissue; R_B is the resistance of the bath and bath

electrode. The dashed line represents shielding and

all of the C's are stray capacitances that may or may

not conduct artifactual current flows.

spectrum will not be flat but will show random variations such as

those shown in the right hand side of Panel D. One must

therefore experimentally measure the input power spectrum, and

perform a few averages in order to ensure there is adequate input

energy at each frequency of interest. Because most experiments

produce additive noise on the record of membrane current or

voltage, a certain amount of averaging is usually preferable.

The disadvantage of using a random input is that the averaging

must be done on the computed power spectra whereas a pseudo

random input is periodic, so the averaging can be done in the

time domain. Moreover, measurement of the input power spectrum

is not essential when using a pseudo random input. However if

one is inherently a skeptic, a knowledge of exactly what signal
is reaching the membrane is reassuring.

Clearly, for frequency domain analysis a pseudo random or
random signal is the extrinsic input of choice. Each has some
advantages and disadvantages over the other, but both have
enormous advantages over more classical inputs. In
microelectrode work the signal to noise ratio is notoriously
poor, so extensive averaging is essential. Moreover, because of
the non-ideal characteristics of a microelectrode, one can never
be sure of the input power spectrum without directly measuring
it. Hence, for microelectrode work a random input has no
limiting disadvantages and in the rest of this paper we will be
considering microelectrode or semi-micro patch electrode
recordings, and the input function will be a random signal.

METHODS

Measurement Artifacts

The utility of studying membrane properties in the frequency
domain lies in the exquisite sensitivity of frequency domain data
to the paths of current flow through a tissue or across a
membrane. This very sensitivity, which makes the technique
valuable, makes the data exceedingly prone to artifact. Figure 2
illustrates the main sources of artifact in a microelectrode
experiment: each is represented by a stray capacitance which may
conduct current.

Whether one measures the power spectrum of intrinsic noise
or the frequency domain response to extrinsic noise, the voltage
recording system will tend to distort the record because of three
stray capacitances. First, the recording amplifier will have an

input capacitance to ground C_{IN} (typically 5 pF); the shielding
of cables or electrodes will introduce a second capacitance C_{SE}
between the signal path and shield (typically 30 pF per foot).
The third source is the glass wall of the electrode which
introduces a stray capacitance between the interior of the
electrode and the bathing solution of about 3 pF per millimeter
of immersion.

One of these capacitances, or sometimes the parallel sum of
two or three, forms a time constant with the tip resistance of
the electrode. This slows down the record of membrane voltage
and introduces phase shift and attenuation in high frequency
data. The potential seriousness of these stray capacitances can
be appreciated by computing the artifactual phase shift at 1 kHz,
for an electrode of 10 MΩ tip resistance, and for a cable six
inches in length, assuming the capacitances C_{IN} and C_{SE} are
uncompensated. The result is greater than 50° of additive
artifactual phase shift at a frequency where the impedance of
most preparations would have a similar amount of phase shift.
Because of the potential magnitude of this error, one must
compensate for these capacitances in some manner, and the
compensation needs to be nearly perfect.

The second source of artifact exists when we apply an
exrinsic input to a preparation, then there will always be some
coupling capacitance C_c between the current passing electrode and
the voltage recording electrode. This capacitance is through the
air between the electrodes, thus, as one might anticipate, it is
quite small and in an unshielded environment will be on the order
of 0.01 pF. Nonetheless, it adds to the recorded impedance a
term given by $R_I R_e j\omega C_c$, which because of the product $R_I R_e$, will
dominate phase data at frequencies around 100 Hz. Thus, a
grounded shield must be placed between the two electrodes,
otherwise impedance studies of most cells or tissues are not

possible. Probably, the most effective method of removing C_c is
to paint both electrodes with conductive silver paint, which is
grounded and then insulated from the bath by overpainting with
varnish. This technique reduces C_c to something sufficiently
small that at a frequency of 10 kHz we cannot detect its
presence.

When applying extrinsic current to a preparation, we always
have a third source of stray capacitance C_I between the interior
of the current passing electrode and the bathing solution. This
capacitance introduces two separate artifacts into impedance
data. First, the usual method of measuring the applied current
is to collect all of the current flowing in the bath, but because
of C_I not all of the bath current is flowing through Z_m. C_I
shunts a significant amount of current around the preparation and
directly into the bath, hence we tend to overestimate the
stimulating current. Second, because the current flowing through
C_I is proportional to frequency, the bath current can become
quite large. Any bath will have some finite resistance, so a
voltage will develop in the bath that is given by $I_{Bath}R_{Bath}$. As
the current into the bath through C_I increases with frequency, so
does the voltage in the bath and our record of the voltage drop
across Z_m becomes contaminated.

The error in recording the membrane voltage is reasonably
easy to cope with. By placing a differential recording electrode
in the bath, the bath voltage may be subtracted from the
intracellular voltage thereby giving an accurate measure of the
transmembrane voltage. The undesirable bath voltage as well as
the error in measurement of membrane current can be minimized by
carefully painting the current electrode. As mentioned above, we
find the glass wall of the microelectrode contributes about 3 pF
of stray capacitance per millimeter of immersion, so by painting
to within 100 μm of the tip, the value of C_I will be reduced to

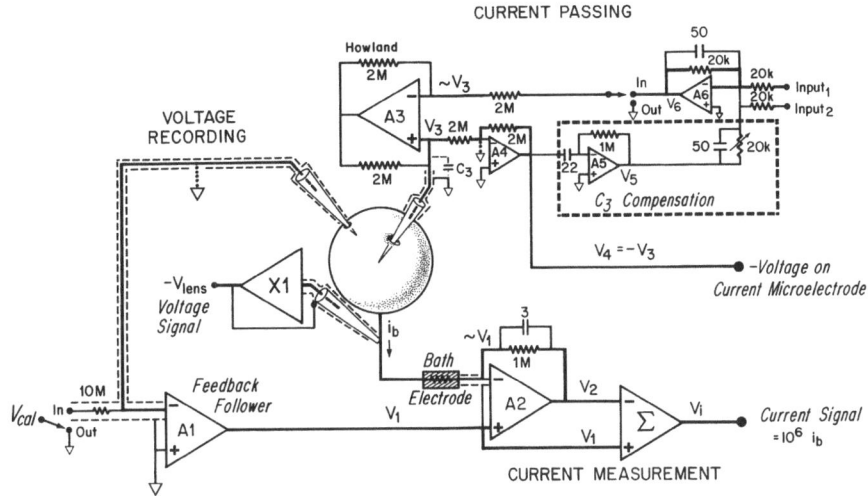

Fig. 3.   From Mathias et al. (1979).   Circuitry used to record
the impedance of the lens.

0.3 pF.   Nevertheless, at frequencies above 1 kHz this small
residual value of C_I shunts enough current around Z_m to preclude
an accurate measurement of the phase angle of Z_m.   The method of
recording impedance data at frequencies above 1 kHz is discussed
in the following section.

Microelectrode Interface Circuits

Many circuits have been designed and used for intracellular
recording or current passing (Valdiosera et al., 1974; Kootsey
and Johnson, 1973; Mathias et al., 1981).   There is no single
"correct" circuit for interfacing with a microelectrode but those
illustrated in Fig. 3 have been proven to work, and moreover, for
the reasons listed below, they are experimentally quite
convenient.   Fig. 3 is from Mathias et al., (1981) where the
operation and mechanization of each circuit is described in

detail. Because these circuits are not unique and because they
are well described elsewhere, in this paper only the reasons for
choosing these particular circuits will be discussed. Hopefully,
if the reader understands why a particular design has been
chosen, then he can choose an implementation to meet his own
needs.

The feedback follower amplifier A1 is the basic membrane
voltage recording stage. The amplifiers labeled "current
measurement" are not an essential component of the voltage
recording operation, and the feedback follower would perform if
the output of A1 were connected directly to the bath electrode.
The preparation is in the feedback path of A1 and, because of the
large input resistance of a FET input op amp, there will be
essentially no current flow through the tip of the voltage
electrode. Hence, the feedback follower is a unity gain device;
it does not change gain if the resistance of the electrode
changes, and it does not draw any measurable current from the
preparation. Moreover, the interior of the voltage recording
electrode as well as any cables between the electrode and
amplifier are driven to virtual ground potential. Thus, the
capacitances C_{IN} and C_{SE} in Fig. 2 have almost no voltage across
them, and the stray current flow through these capacitors is
reduced to a negligible amount. If the voltage electrode is
carefully painted, then the dominant source of artifact is the
small residual capacitance C_e between the electrode interior and
the bath. The residual time delay introduced by the time
constant $R_e C_e$ is easily measured by applying a calibration input
V_{cal} and recording V_1. We may therefore record a calibration
curve for each electrode and in subsequent data analysis
essentially remove all remaining errors.

In order to subtract the voltage drop through the resistance
of the bath electrode, we place a low resistance glass pipette in

the bath, near to the preparation, and record the differential
bath voltage with a standard follower circuit (Smith, 1971). The
only precaution necessary with low resistance pipettes (<500 kΩ)
is to drive the shielding to remove the relatively large
capacitance associated with shielded cables. The differential
circuit measures the bath voltage with a time delay due to the
input capacitance of the op amp times the pipette resistance. If
we apply a calibration input V_{cal} to A1, and record the response
through the differential circuit, then we will record the delays
of both feedback follower and differential electrode, thus one
calibration curve will suffice to remove all residual errors in
the record of membrane voltage. With these two circuits
implemented as described, one can accurately determine the
transmembrane voltage to at least 10 kHz.

The two "current measurement" amplifiers record all of the
current flowing in the bath. Although this current can be
accurately measured to tens of kHz, the measurement is only
meaningful to about 1 kHz, due to the shunting effect of C_I
(Fig. 2) discussed in the previous section.

The current passing amplifier is essentially a Howland
constant current source (Smith, 1971). We choose to use a
constant current circuit because of a problem that has not yet
been described, but is always present when using microelectrodes.
The tip resistance of a microelectrode is not an ideal circuit
element. Indeed, it tends to rectify or pass current more easily
in one direction than the other, and it tends to pass high
frequency current more easily than low frequency current. If a
voltage with a flat power spectrum is applied on top of a
microelectrode, the current which results will not have a flat
power spectrum. Very little energy will exist at low frequencies
in the current signal, thus, if we use this current to stimulate
a membrane, the signal to noise ratio for the low frequency

components of the membrane voltage will be unnecessarily poor. A
constant current circuit will accurately control the low
frequency components of current and thereby ensure an improved
and more uniform signal to noise ratio.

The problem in using constant current circuits with
microelectrodes is that total current is controlled, not just the
microelectrode tip current. At high frequencies most of the
current flows through the stray capacitance C_{SI} illustrated in
Figure 2; hence we get a poor signal to noise ratio at high
frequencies. The three amplifiers other than A3 in the "current
passing" circuit increase the high frequency voltage command to
the Howland, thus the current lost to stray capacitance at high
frequencies is compensated by the extra command voltage at high
frequency. In this configuration the Howland will deliver
sufficient current to the preparation at frequencies well above
10 kHz. At frequencies above 1 kHz, however, we cannot
accurately measure the current flowing through the electrode tip
resistance and into the preparation, so another means of
recording the impedance must be employed.

The tip of a glass electrode is well modeled by a parallel
resistor and capacitor (Valdiosera et al., 1974). If a voltage
is applied across a parallel resistor and capacitor, the current
will divide into two components: the component flowing through
the resistor will be in phase with the voltage, whereas the
component flowing through the capacitor is 90° out of phase with
the voltage. The current flowing through the tip resistance of
the electrode will enter the preparation, whereas the current
through the capacitance of the electrode wall is mostly shunted
into the bath. The voltage on top of the current electrode
therefore provides a phase reference from which the phase angle
of the membrane impedance can be measured. Thus, at frequencies
above 1 kHz we can accurately measure the phase angle of the

impedance but not the magnitude, whereas at frequencies below 1 kHz we can accurately measure both magnitude and phase. Fortunately, the phase angle at all frequencies and the magnitude at just one frequency are sufficient to completely specify a strictly stable, physically realizable transfer function (Zadeh and Desoer, 1963) so we have more than enough data to determine the parameter values for the equivalent circuit of the preparation.

The most difficult part of a microelectrode impedance experiment is obtaining high fidelity time domain data. Nevertheless, there are a number of steps between recording data and interpreting data, and each step may introduce some new artifact. Before analysis of data can begin, one must store a finite set of data points. This is most usually done by passing the time domain functions through an analog to digital (A/D) converter and storing a representative number of points in a digital computer. A potentially serious source of artifact during A/D conversion is the phenomenon of aliasing.

Aliasing

When we perform an A/D conversion, the continuous time function will be sampled at some finite rate. The sample rate sets the bandwidth over which the frequency components of the function can be resolved. If the time function contains sinusoidal components at frequencies greater than one half the sample frequency, we cannot detect these components, yet they can have a devastating effect on our sampled data. Figure 4 illustrates what can happen if we sample at 5 Hz in the presence of a 4 Hz signal. The 4 Hz signal produces a set of data points that describe a 1 Hz sine wave, even though there may be no actual 1 Hz component present in the sampled function.

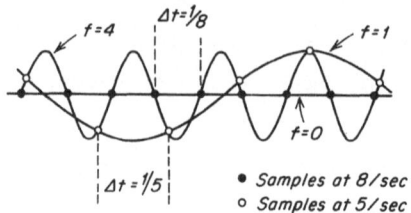

Fig. 4.   From Eisenberg (1983). The phenomenon of aliasing. The
          actual signal is at a frequency f = 4 Hz, but because
          the sample rate is only at 5 Hz, the apparent
          (digitized) signal is at f = 1 Hz.

        High frequency, unresolvable components of a signal are
aliased back as apparent low frequency components; the first
order aliasing is symmetrical about what is called the folding
frequency. The folding frequency equals one half of the sample
frequency, thus, in Figure 4 where the sample frequency is 5 Hz,
the folding frequency is 2.5 Hz and frequency components equal
distant from 2.5 Hz are principle aliases of one another. In
Figure 4 the high frequency component is at 4 Hz, which is 1.5
above the folding frequency; a 4 Hz signal folds back to 1.5 Hz
below the folding frequency, which is 1 Hz. The 4 Hz component
is the principle alias of 1 Hz, but components at all multiples
of 4 Hz will also be aliased back as a 1 Hz component in the
sampled data.

        The method of reducing this artifact is straightforward.
One must perform analog filtering prior to A/D conversion thereby
reducing the amplitude of the aliased components to an acceptable
level. In theory, one should remove all components at
frequencies greater than one half the sample rate. In practice,
analog filters have a specified rate of attenuation with
frequency, so in order to achieve a desired level of attenuation

at one half the sample frequency, some potentially useable data
will be distorted by the anti-aliasing filter. Hence, it is not
feasible to accurately recover frequency components at the
theoretical limit. However, the steeper the roll off rate of the
filter, the more useable data we can recover.

A numerical example may clarify the above discussion.
Assume we are willing to accept aliasing of components attenuated
by more than 0.001, and assume we have an analog filter which
does not significantly distort frequency components below 1 Hz,
but attenuates components at frequencies above 10 Hz by more than
0.001. Accordingly, we are willing to allow the 10 Hz signal to
be aliased back as a 1 Hz signal, thus the minimum folding
frequency should be 5.5 Hz and our sample rate must be at least
11 Hz. In a 10 second time period we will record 110 data
points, but we can only resolve signals of 1 Hz and below, thus
only 2 out of every 11 points are used to recover the undistorted
sinusoidal components of the time signal.

Thus far, the methods of recording and digitizing high
fidelity time domain data have been discussed. Because analysis
will ultimately be accomplished in the frequency domain, the time
domain points must be converted into their frequency domain
counterparts. If a pseudo random periodic input were used, then
the time domain response may be averaged, the average response
can be digitized and represented as a Fourier series by standard
digital Fourier transform software, and equations (2) through (5)
will accurately define the impedance function $Z(j\omega)$. If a zero
mean random input is used, then time domain signal averaging will
simply yield zero and is not useful. The averaging must
therefore be accomplished after the data are digitized and
Fourier transformed. Although one can conceive of several
schemes for averaging frequency domain functions, most schemes
will yield a biased estimate of $Z(j\omega)$ (Bendat and Piersol, 1971).

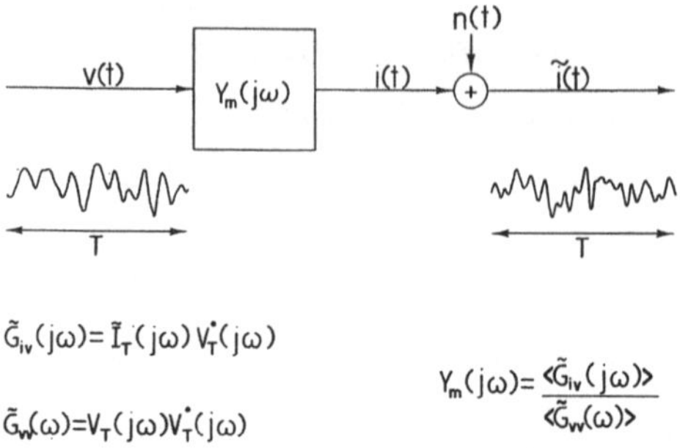

$$\tilde{G}_{iv}(j\omega)= \tilde{I}_T(j\omega)V_T^*(j\omega)$$

$$\tilde{G}_w(\omega)=V_T(j\omega)V_T^*(j\omega)$$

$$Y_m(j\omega)= \frac{\langle\tilde{G}_{iv}(j\omega)\rangle}{\langle\tilde{G}_w(\omega)\rangle}$$

Fig. 5.   Estimation of the admittance Y_m using an extrinsic
          voltage noise V(t) as an input and recording current
          i(t) in the presence of additive, contaminating noise
          n(t). The equations show the method of averaging < >
          which removes the effect of n(t) from our computation of
          Y_m. The Fourier transforms subscripted with T mean that
          a time record of length T was transformed. The
          averaging is over many transforms of different time
          records, each of length T.

The following method produces an unbiased estimate of $Z(j\omega)$, and
it is the method used in most commercially available Fourier
analyzers.

Computation of a Transfer Function

    Figure 5 is a block diagram of a voltage clamp experiment.
The transfer function in this experiment is the membrane
admittance, $Y_m(j\omega)$.   (Note $Y_m(j\omega) = 1/Z_m(j\omega)$).   A voltage of
random time course v(t) is the input signal; the output signal is
the current i(t). In order to clamp the membrane to the
potential v(t), i(t) is automatically supplied by some feedback
network, but because of contamination from intrinsic membrane
noise, instrumentation noise, and Johnson noise from the

electrodes, the current $i(t)$ cannot be accurately measured.
Hence, the measured current is $\tilde{i}(t)$:

$$\tilde{i}(t) = i(t) + n(t)$$

where $n(t)$ is additive noise from the sources described above.

In theory, the transfer function can be computed from the
cross correlation function $r_{iv}(t)$ and voltage auto correlation
function $r_{vv}(t)$. Define:

$$r_{iv}(t) \triangleq \lim_{T\to\infty} \frac{1}{2T} \int_{-T}^{T} \tilde{i}(t+\tau)v(\tau)d\tau \qquad (6)$$

$$r_{vv}(t) \triangleq \lim_{T\to\infty} \frac{1}{2T} \int_{-T}^{T} v(t+\tau)v(\tau)d\tau \qquad (7)$$

Fourier transforming these functions yields

$$G_{IV}(j\omega) \triangleq \int_{-\infty}^{\infty} e^{-j\omega t} r_{iv}(t)dt \qquad (8)$$

$$G_{VV}(\omega) \triangleq \int_{-\infty}^{\infty} e^{-j\omega t} r_{vv}(t)dt \qquad (9)$$

where $G_{VV}(\omega)$ is the power spectrum of the input voltage and
$G_{IV}(j\omega)$ is the cross power spectrum for the output current with
input voltage. The admittance $Y_m(j\omega)$ is given by

$$Y_m(j\omega) = G_{IV}(j\omega)/G_{VV}(\omega) \qquad (10)$$

Equations (6) through (10) are well known theoretical results (Papoulis, 1965), nonetheless, to an experimentally oriented scientist, they probably appear somewhat circuitous. For example, why not Fourier transform the output and input signals, and compute the transfer function from the ratio? The reason is in part a theoretical problem, but the practical problem of distinguishing output current noise i(t) from additive measurement noise n(t) also demands circuitous methods.

The theoretical problem arises from a typical approximation: the theory assumes the input noise has been applied since $t = -\infty$. Both the output current and input voltage therefore, in theory, contain infinite energy, and such signals do not possess Fourier transforms. In an experiment, the input and output signals can exist only over a finite time period and must contain finite energy, hence we can take Fourier transforms with impunity. If we record the input and output signals over the time $o \le t \le T$, then each signal can be Fourier transformed

$$V_T(j\omega) = \int_o^T e^{-j\omega t} v(t) dt \tag{11}$$

$$\tilde{I}_T(j\omega) = \int_o^T e^{-j\omega t} \tilde{i}(t) dt \tag{12}$$

An estimate of the cross power spectrum and input power spectrum may be computed from

$$\tilde{G}_{IV}(j\omega) = \tilde{I}_T(j\omega) V_T^*(j\omega) \tag{13}$$

$$\tilde{G}_{VV}(\omega) = V_T(j\omega) V_T^*(j\omega) \tag{14}$$

where * indicates complex conjugation and the tilde ~ indicates
an estimate.

If we record $\tilde{i}(t)$ and $v(t)$ over a new time period of length
T, then we may average our second estimates with our first
estimates, and so on, until we obtain arbitrarily good estimates
of the spectra. If we denote the process of taking an infinite
number of averages (i.e. the mathematical "expected value")
(Papoulis, 1965) by brackets < >, then

$$G_{IV}(j\omega) = <\tilde{G}_{IV}(j\omega)>$$

$$G_{VV}(\omega) = <\tilde{G}_{VV}(\omega)> \qquad (15)$$

and

$$Y_m(j\omega) = \frac{<\tilde{G}_{IV}(j\omega)>}{<\tilde{G}_{VV}(\omega)>} \qquad (16)$$

It is important to note that the average value for $Y_m(j\omega)$ is not
obtained by averaging estimates of the admittance, rather it is
obtained by averaging estimates of the spectra, then taking the
ratio of averaged cross spectrum to average input spectrum. It
can be shown (Bendat and Piersol, 1971) that if estimates of the
transfer function (admittance) are averaged, the expected value
of the average will be biased, (i.e. the average of the ratio is
not the ratio of the averages).

Moreover, it is important to note that the effect of
additive noise on the output is removed by this scheme of
averaging. Because n(t) and v(t) are uncorrelated, their Fourier
transforms are also uncorrelated. The average of a sum is the
sum of the averages, hence

$$\langle \tilde{G}_{IV}(j\omega)\rangle = \langle V_T^*(j\omega)I_T(j\omega)\rangle + \langle V_T^*(j\omega)N_T(j\omega)\rangle \tag{17}$$

but

$$\langle V_T^*(j\omega)N_T(j\omega)\rangle = 0 \tag{18}$$

since they are uncorrelated.

If one were to compute the power spectrum of the current, there would be a bias introduced by n(t). Viz.:

$$\langle \tilde{G}_{II}(\omega)\rangle = \langle |I_T(j\omega) + N_T(j\omega)|^2\rangle = \langle |I_T j\omega)|^2\rangle + \langle |N_T(j\omega)|^2\rangle$$

$$= G_{II}(\omega) + G_{NN}(\omega) \tag{19}$$

Two considerations arise from this observation. One, when computing the membrane admittance or impedance, the noisy signal should always be considered the output. Otherwise the estimate of the input power spectrum will be biased. And second, by applying an extrinsic input and observing the correlated response, we have avoided a serious problem in fluctuation analysis: separation of channel noise from background noise. In fluctuation analysis one simply measures the power spectrum of the membrane current and this will be contaminated by the power spectrum of additive noise.

ANALYSIS OF CHANNEL PROPERTIES

Extrinsic noise can be used to study properties of channels in much the same manner as intrinsic noise. In either approach, one records frequency domain data and compares these data with a theoretical model of channel gating, and so both approaches give results which are model dependent. The basic kinetic model in

either study is the same, yet each experimental protocol
emphasizes a different aspect of the model, thus performing both
protocols will help to discriminate between models. Moreover,
each experiment has unique sources of artifact. For example,
background noise will contaminate measurement of the power
spectrum of intrinsic noise, whereas analysis of the response to
extrinsic noise assumes a linear relationship between membrane
current and voltage. The effects of these potential sources of
error need to be evaluated. Because this paper is concerned with
extrinsic noise, some practical aspects of errors introduced by
nonlinearities will be discussed; errors inherent in the
recording and analysis of intrinsic noise should be discussed
elsewhere in this book.

The most simple linear circuit is a resistor, where current
and voltage are related by a multiplicative constant,
$(v(t) = R\, i(t))$. In the more usual situation, the current and
voltage are related through a differential equation, and direct
proportionality will not hold. That is, memory is introduced
because of the differential equation, so the current at time t
not only depends on the voltage at time t, but also depends on
the voltage at all times prior to t. Nonetheless, if one defines
some measure of the total input current, for example the energy
contained in the input signal, then in a linear system the energy
in the output will be directly proportional to the energy in the
input, if there are no energy sources in the circuit. Hence, a
time independent resistor (i.e. a system where input and output
are proportional) can provide useful insight into linear behavior
and moreover, into the effects of nonlinearities.

Nonlinear Effects

A simple nonlinear circuit is a resistor which depends on
current; for example let

$$v(t) = r_1 i(t) + r_2 i^2(t) \tag{20}$$

In general, a real resistance will increase with increasing temperature, and temperature will increase in proportion to power dissipated. Hence, equation (20) is an approximate model of most resistors, but because r_2 is a small parameter, we generally treat resistors as linear circuit elements. Indeed, almost nothing is truly linear, yet many naturally occurring systems are successfully analyzed using linear theory.

In many situations, r_2, the coefficient of the first nonlinear effect, is a small number. This is analogous to studying excitable membranes outside of the range of voltages where nonlinear conductances are active (Cole, 1972; Mathias et al., 1979; Falk and Fatt, 1964; Levis et al., 1983). In other circumstances, r_2 may be comparable to r_1 but if we keep the current levels small, such that $i^2(t) << i(t)$, then we can still approximate the resistance as being linear (Cole, 1972; Mathias et al., 1981; Fishman et al., 1981). Indeed, for any values of r_1 and r_2, one can always apply a sufficiently small value of $i(t)$ such that $r_2 i^2(t) << r_1 i(t)$. The effects of nonlinearity are always dependent on the size of the input signal, so we can, in practice, almost always study the linear behavior of a system by keeping the input signal sufficiently small.

The Volterra Theory of Nonlinear Systems

Consider the simple nonlinear current voltage relationship described by equation (20). If we chose to apply the voltage as the input and record the current response, then we must revert the sum in (20) (Abramowitz and Stegun, 1972), and we find the current voltage relationship requires an infinite series.

$$i(t) = g_1 v(t) + g_2 v^2(t) + g_3 v^3(t) + \ldots \tag{21}$$

where

$$g_1 = 1/r_1 \; ; \quad g_2 = -r_2/r_1^3 \; ; \quad g_3 = 2r_2^2/r_1^5 \; ; \quad \ldots \tag{22}$$

The Taylor series in equation (21) is a special case of the Volterra theory of nonlinear systems (Schetzen, 1980). In the more general case, current and voltage are related by a nonlinear differential equation, and the coefficients g_1, g_2, ... represent operators on the functions $v(t)$, $v^2(t)$, ... hence the general Volterra series is simply a Taylor series with memory.

The utility of Volterra's approach is that the coefficients (or operators) g_1, g_2, ... define the nonlinear circuit, regardless of the time course of the input. Moreover, the nth component of the series is pseudo linear in $v^n(t)$, in the sense that multiplying $v(t)$ by ε, results in multiplication of the nth component by ε^n. The first term in the series is therefore the linear response of the system, (note $g_1 = 1/r_1$ in equation (22)). The problem with this approach is that for arbitrarily large inputs, the series converges very slowly or not at all. But for the purposes of this discussion, we will assume the input is sufficiently small that a linear approximation is good, and the next order correction is adequate to assess errors introduced by nonlinearities.

If a system is linear, then a sinuosidal input of frequency ω_1 will produce a sinusoidal output, also at the frequency ω_1. The effect of nonlinearities in the system is to spread the frequencies over which the response will occur. When the circuit described by equation (21) is driven by a sinusoidal input voltage,

$$v(t) = \varepsilon e^{j\omega_1 t} \tag{23}$$

the output current will contain sinusoids whose frequencies are integer multiples of ω_1.

$$i(t) = \epsilon g_1 e^{j\omega_1 t} + \epsilon^2 g_2 e^{2j\omega_1 t} + \epsilon^3 g_3 e^{3j\omega_1 t} + \ldots \tag{24}$$

If the input voltage were two sinusoids at frequencies ω_1 and ω_2, then the output current will contain linear combinations of the input frequencies.

$$v(t) = \epsilon(e^{j\omega_1 t} + e^{j\omega_2 t}) \tag{25}$$

$$i(t) = \epsilon g_1 (e^{j\omega_1 t} + e^{j\omega_2 t})$$

$$+ \epsilon^2 g_2 (e^{2j\omega_1 t} + 2e^{j(\omega_1 + \omega_2)t} + e^{2j\omega_2 t})$$

$$+ \epsilon^3 g_3 (e^{3j\omega t} + 3e^{j(2\omega_1 + \omega_2)t} + 3e^{j(\omega_1 + 2\omega_2)t} + e^{3j\omega_2 t})$$

$$+ \ldots \tag{26}$$

In the more general case, where the circuit is described by a nonlinear differential equation, the multiplicative constants g_n are replaced by operators,

$$G_n(v^n(t)) = \int_{-\infty}^{\infty} \int_{-\infty}^{\infty} \ldots \int_{-\infty}^{\infty} y_n(t_1, t_2, \ldots, t_n) v(t - t_1)$$

$$\ldots v(t - t_n) dt_1 \ldots dt_n$$

In this circumstance, when the input voltage is described by equation (25), the output current is

$$i(t) = \epsilon \int_{-\infty}^{\infty} y_1(\tau)[e^{j\omega_1(t - \tau)} + e^{j\omega_2(t - \tau)}]d\tau$$

$$+ \epsilon^2 \int_{-\infty}^{\infty} \int_{-\infty}^{\infty} y_2(\tau,s)[e^{j\omega_1(t - \tau)} + e^{j\omega_2(t - \tau)}][e^{j\omega_1(t - s)}$$

$$+ e^{j\omega_2(t - s)}]d\tau ds + \dots \tag{27}$$

or

$$v(t) = \epsilon[Y_1(j\omega_1)e^{j\omega_1 t} + Y_1(j\omega_2)e^{j\omega_2 t}]$$

$$+ \epsilon^2[Y_2(j\omega_1,j\omega_1)e^{j\omega_1 t} + 2 Y_2(j\omega_1,j\omega_2)e^{j(\omega_1 + \omega_2)t}$$

$$+ Y_2(j\omega_2,j\omega_2)e^{2j\omega_2 t}] + \dots \tag{28}$$

where $Y_1(j\omega)$ is the linear admittance of the circuit and, $Y_2(j\omega_1, j\omega_2)$ is the two dimensional Fourier transform of the second order Volterra kernel, given by

$$Y_2(j\omega_1,j\omega_2) = \int_{-\infty}^{\infty} \int_{-\infty}^{\infty} y_2(\tau_1,\tau_2)e^{-(j\omega_1\tau_1 + j\omega_2\tau_2)}d\tau_1 d\tau_2 \tag{29}$$

It is easily shown (Schetzen, 1980) that

$$Y_2(j\omega_1,j\omega_2) = Y_2(j\omega_2,j\omega_1)$$

consequently, the cross frequency terms in equation (27) combine to give $2Y(j\omega_1,j\omega_2)$ in equation (28).

In equation (28) the second order response is the sum over all permutations of the input frequencies ω_1 and ω_2 taken two at a time. If the input voltage contained N sinusoidal components, then it is easy to see that the second order response would contain all permutations of N input frequencies taken two at a time. The extension of this result to arbitrary input waveforms is an integral of all permutations taken two at a time.

For example, if v(t) is an arbitrary physically realizable input whose Fourier transform is $V(j\omega)$,

$$V(j\omega) = \int_{-\infty}^{\infty} e^{-j\omega t} v(t)dt,$$

and i(t) is the response of a nonlinear differential equation to the small input function $\varepsilon v(t)$, then it can be shown (Schetzen, 1980) that

$$I(j\omega) = \varepsilon Y_1(j\omega)V(j\omega)$$

$$+ \varepsilon^2 \frac{1}{2\pi} \int_{-\infty}^{\infty} Y_2(j\Omega, j\omega - j\Omega)V(j\Omega)V(j\omega - j\Omega)d\Omega + \ldots \quad (30)$$

Equation (30) illustrates the inherent complexity of dealing with any nonlinear system in a reasonably general fashion, but it also illustrates a method of rigorously evaluating the range of ε over which a system can be considered linear.

The equations presented in this section are most useful if one has some idea of the form of the nonlinear differential equation describing the circuit of interest. In other words,

this presentaiton has focused on system analysis rather than
system identification. In problems of system identification, the
kernels are unknown functions which need to be synthesized,
whereas in system analysis the form of the kernels is thought to
be known, but the importance or value of parameters in the system
needs to be analyzed. System analysis is most efficiently
accomplished by linear frequency domain techniques, therefore the
Volterra series is the most appropriate representation of the
system, since it allows linear analysis plus estimation of errors
due to nonlinearities. If one is required to confront the
problem of system identification, then the more general Wiener
series representation of a nonlinear system is appropriate
(Wiener, 1958; Marmarelis and Naka, 1974). Wiener's theory
however is not a panacea of nonlinear systems, and many of the
possible pitfalls are pointed out by Palm and Poggio (1977a,b).

Volterra Series for a First Order Channel

The theoretical framework derived by Volterra, and outlined
in the preceding paragraphs, can be used to analyze the gating of
biological channels. However, we must first propose a model of
channel gating, then we may determine its linear response, and
the second order correction term to a linear analysis.

One simple model of channel gating was proposed by Hodgkin
and Huxley. They hypothesized a channel as an aqueous pore that
can be closed by one or more gates, and each gate opens or closes
via a first order chemical transition. The rate constants for
the transition are assumed to be functions of membrane voltage,
thus the current voltage relationship for the membrane will be
nonlinear, but the equation describing the probability of a
transition is linear. If there is only one first order gate in a
channel, then the channel will be called "first order". Although
this model is certainly too simple to exactly describe any

channel, it has approximately described many channels and is
therefore a suitable archetype for the purpose of this analysis.

When analyzing biological membrane currents, one must always
keep in mind that each current has a different driving force or
equilibrium potential. The presence of these equilibrium
potentials can introduce currents that are proportional to
conductance fluctuations, so we must keep an account of each
equilibrium voltage. Assume the membrane has a resting potential
E_R when $g(v,t) = 0$, and that the channel has an equilibrium
potential E, then if we voltage clamp the membrane to a voltage
$v(t)$ we have

$$i(t) = G_m[v(t) - E_R] + C_m \frac{dv(t)}{dt} + g(v,t)[v(t) - E] \qquad (31)$$

where G_m and C_m represent the linear, voltage and time
independent components of the membrane conductance and
capacitance, and $g(v(t),t)$ is the conductance contributed by the
ensemble of first order channels within the membrane. The
dependence of $g(v(t),t)$ on $v(t)$ is functional in nature and is
most often described implicitly by a differential equation. The
differential equation is based on the probability $p(v(t),t)$ that
a channel is in the open state.

$$1 - p(v) \underset{\beta(v)}{\overset{\alpha(v)}{\rightleftarrows}} p(v) \qquad (32)$$

closed   open
state   state

where $\alpha(v)$ and $\beta(v)$ are the voltage dependent rate constants that
represent the probability per unit time of a change in state.
The conductance $g(v,t)$ is given by the conductance $\bar{g}$, which is

the maximal conductance when all of the channels are open,
multiplied by the probability function $p(v,t)$, which is the
fraction of open channels at the voltage v and time t.

$$g(v,t) = \bar{g}p(v,t) \tag{33}$$

Equation (32) is a pictorial representation of the differential
equation that describes $p(v,t)$.

$$\frac{dp(v,t)}{dt} = -\omega_o(v)[p(v,t) - P_o(v)] \tag{34}$$

where

$$\omega_o(v) = \alpha(v) + \beta(v) \tag{35}$$

$$P_o(v) = \alpha(v)/[\alpha(v) + \beta(v)] \tag{36}$$

If the input voltage $v(t)$ is given by a steady value $\bar{V}$ and a
small time varying component $\varepsilon v_m(t)$, then equations (31) and (34)
can be expanded in a Volterra series. Viz.,

$$v(t) = \bar{V} + \varepsilon v_m(t) \tag{37}$$

where

$$|v_m(t)| \le 1, \quad |\varepsilon| < 1 \tag{38}$$

then,

$$i = i_o + \varepsilon i_1 + \varepsilon^2 i_2 + \ldots$$
$$\tag{39}$$
$$g = g_0 + \varepsilon g_1 + \varepsilon^2 g_2 + \ldots$$

Substituting (39) into (31) gives

$$i_0 + \varepsilon i_1 + \varepsilon^2 i_2 + \ldots = G_m(\bar{V} - E_R) + g_0(\bar{V} - E)\varepsilon[(G_m + g_0)v_m$$

$$+ C_m(dv_m/dt) + g_1(\bar{V} - E)]$$

$$+ \varepsilon^2[g_1 v_m + g_2(\bar{V} - E)] + \ldots$$

Collecting terms of like order in ε yields

$$i_0 = G_m(\bar{V} - E_R) + g_0(\bar{V} - E) \qquad \text{(Baseline)}$$

$$i_1 = (G_m + g_0)v_m + C_m\frac{dv_m}{dt} + g_1(\bar{V} - E) \quad \text{(Linear response)} \quad (40)$$

$$i_2 = g_1 v_m + g_2(\bar{V} - E) \qquad \text{(Volterra correction)}$$

In order to determine g_0, g_1, g_2, $\ldots$, we substitute the expansion of g given in equation (39) into the differential equation (34) for p $(= g/\bar{g})$. Equation (34) can be further expanded in orders of ε by performing a Taylor expansion of $\omega_0(v)$, $P_0(v)$ about the point $v = \bar{V}$. Viz.:

$$\omega_0 = \bar{\omega}_0 + \varepsilon v_m \bar{\omega}_1 + \varepsilon^2 v_m^2 \bar{\omega}_2 + \ldots$$

where   (41)

$$\bar{\omega}_k = \frac{1}{k!} \frac{\partial^k \omega_0(v)}{\partial v^k}\Bigg|_{v = \bar{V}}$$

and

$$P_0 = \bar{P}_0 + \varepsilon v_m \bar{P}_1 + \varepsilon^2 v_m^2 \bar{P}_2 + \dots$$

where $\qquad\qquad\qquad\qquad\qquad\qquad\qquad\qquad\qquad\qquad$ (42)

$$\bar{P}_k = \frac{1}{k!} \left. \frac{\partial^k P_0(v)}{\partial v^k} \right|_{v = \bar{v}}$$

and we can define

$$\bar{G}_k = \bar{g}\bar{P}_k \qquad\qquad\qquad\qquad\qquad (43)$$

Hence,

$$\frac{d}{dt}(g_0 + \varepsilon g_1 + \dots) = -[\bar{\omega}_0 + \varepsilon v_m \omega_1 + \dots][(g_0 - \bar{G}_0)$$

$$+ \varepsilon(g_1 - \bar{G}_1 v_m) + \dots]$$

Once again collecting terms of like powers in ε, we obtain the following equations and solutions.

$$g_0 = \bar{G}_0 \qquad\qquad\qquad\qquad\qquad (44)$$

$$\frac{dg_1}{dt} = -\bar{\omega}_0[g_1 - \bar{G}_1 v_m]$$

$$\qquad\qquad\qquad\qquad\qquad\qquad\qquad\qquad (45)$$

$$g_1(t) = \bar{\omega}\bar{G}_1 \int_{-\infty}^{\infty} e^{-\bar{\omega}_0(t - \tau)} v_m(\tau) d\tau$$

and

$$\frac{dg_2}{dt} = -\bar{\omega}_0 [g_2 - \bar{G}_2 v_m^2] - \bar{\omega}_1 v_m [g_1 - \bar{G}_1 v_m]$$

(46)

$$g_2 = \int_{-\infty}^{\infty} e^{-\bar{\omega}_0(t-s)} \{\bar{\omega}_0 \bar{G}_2 v_m^2(s) + \bar{\omega}_1 v_m(s) [\bar{G}_1 v_m(s) - g_1(s)]\} ds$$

The generic result is

$$\frac{dg_k}{dt} = -\bar{\omega}_0 [g_k - \bar{G}_k v_m^k] - \bar{\omega}_1 v_m [g_{k-1} - \bar{G}_{k-1} v_m^{k-1}]$$

$$\ldots - \bar{\omega}_{k-1} v_m^{k-1} [g_1 - \bar{G}_1 v_m]$$

$$g_k(t) = \int_{-\infty}^{\infty} e^{-\bar{\omega}_0(t-s)} \{\bar{\omega}_0 \bar{G}_k v_m^k(s) + \bar{\omega}_1 v_m(s) [\bar{G}_{k-1} v_m^{k-1}(s)$$

$$- g_{k-1}(s)] \ldots + \bar{\omega}_{k-1} v_m^{k-1}(s) [\bar{G}_1 v_m(s) - g_1(s)]\} ds \quad (47)$$

The terms in equation (40) have now all been analyzed and we
can determine the "linear response" and "Volterra correction".
In particular, we are interested in the frequency domain
representation of each term. Computation of the Fourier
transform of equation (40) requires us to apply the rule that a
product of two functions in the time domain is equivalent to a
convolution integral in the frequency domain. We will denote the
Fourier transform by upper case functions, then

$$I_1(j\omega) - [G_m + \bar{G}_0 + j\omega C_m + \frac{(V - E)\bar{G}_1}{1 + j\omega/\omega_0}]V_m(j\omega) \quad \text{(Linear} \quad (48)$$
$$\text{Response)}$$

and

$$I_2(j\omega) = \frac{1}{2\pi} \int_{-\infty}^{\infty} \frac{V_m(j\Omega)V_m(j\omega - j\Omega)}{1 + j\Omega/\bar{\omega}_0} \left(\bar{G}_1 + \bar{G}_2[V - E] \frac{1 + j\Omega/\chi\bar{\omega}_0}{1 + j\omega/\bar{\omega}_0} \right) d\Omega$$

$$\text{(Volterra correction)} \quad (49)$$

where

$$\chi = 1/(1 + \bar{G}_1\bar{\omega}_1/\bar{G}_2\bar{\omega}_0) \quad (50)$$

Equation (48) defines the linear membrane admittance $Y_m(j\omega)$ illustrated in Figure 6. The two dimensional transform of the second Volterra kernel is $Y_2(j\omega_1,j\omega_2)$ (S/volt), which can be deduced by comparing equation (49) with equation (30) and substituting $\Omega = \omega_1$, $\omega - \Omega = \omega_2$.

$$Y_2(j\omega_1,j\omega_2) = \frac{1}{1 + j\omega_1/\bar{\omega}_0} \left(\bar{G}_1 + \frac{\bar{G}_2(\bar{V} - E)(1 + j\omega_1/\chi\bar{\omega}_0)}{1 + j(\omega_1+\omega_2)/\bar{\omega}_0} \right) \quad (51)$$

Note that equation (51) is not symmetric in ω_1, ω_2, as was assumed in the earlier analysis. We could easily construct a symmetric form of equation (51) however, because we could have written (49) in the following form:

$$\int_{-\infty}^{\infty} Y_2(j\Omega, j\omega - j\Omega) V_m(j\Omega) V_m(j\omega - j\Omega) d\Omega$$

$$= \int_{-\infty}^{\infty} Y_2(j\omega - j\Omega) V_m(j\omega - j\Omega) V_m(j\Omega) d\Omega = \int_{-\infty}^{\infty} \frac{1}{2} [Y_2(j\Omega, j\omega - j\Omega)$$

$$+ Y_2(j\omega - j\Omega, j\Omega)] V_m(j\Omega) V_m(j\omega - j\Omega) d\Omega$$

The symmetric form of Y_2 will obviously be more complicated than that presented in (51), thus we will try to make the best of both forms by using (51) in explicit results yet assuming Y_2 is symmetric in general results. The multiple forms of Y_2 illustrate its mathematical rather than physical basis and one should always return to equation (49) for physical insight.

Convergence of the Volterra Series and Linearity

In the analysis presented here we have assumed that $|v_m(t)| \leq 1$, hence we can bound the above results by substituting the maximal driving force $v_m(t) = 1$ and computing a bound for each term. From equations (45) and (46), we obtain

$$\max |g_1(v_m, t)| \leq |g_1(1, t)| = |\bar{G}_1| \tag{52}$$

$$\max |g_2(v_m, t)| \leq |g_2(1, t)| = |\bar{G}_2| \tag{53}$$

Moreover, in order to bound $g_2(1, t)$ we see that the second term in the integral equation (46) integrates to zero, i.e.

$$\int_{-\infty}^{\infty} e^{-\bar{\omega}_0(t-s)} [\bar{G}_1 - g_1(1, s)] ds = 0$$

Hence, the equation for $g_2(1,t)$ is identical (with a change of index) to the equation for $g_1(1,t)$. Accordingly, the integral equation for $g_3(1,t)$ does not depend on $[\bar{G}_2 - g_2(1,s)]$ nor does it depend on the term $[\bar{G}_1 - g_1(1,s)]$: it must therefore be identical (with a change of index) to the equation for $g_2(1,t)$. By induction then, we see that

$$\max|g_k(v_m,t)| \leq |g_k(1,t)| = |\bar{G}_k| \tag{54}$$

This interesting result shows the Volterra series for the channel can be termwise bounded by the Taylor series expansion for the steady state behavior. Because the channel obeys a differential equation, it has memory and is inherently more complex than the simple nonlinear conductor analogy given in equation (26). Nonetheless, the Volterra series, which includes the effect of memory, is bounded by the simple nonlinear conductor's Taylor series. This demonstrates the close correspondence between Volterra's approach and a Taylor expansion. Moreover, we can now estimate the value of ε over which we can approximate the channel as being linear. For example, we can neglect the first nonlinear Volterra correction term when

$$|\varepsilon v_m| << |\bar{G}_1|/|\bar{G}_2| \tag{55}$$

These theoretical calculations will perhaps be more meaningful if we apply them to the potassium channel in squid axon, first described by Hodgkin and Huxley (1952). Figure 6 illustrates the voltage dependence of $\bar{G}_0$, $\bar{G}_1$ and $\bar{G}_2$ for the squid K^+ channel, using the parameters from Hodgkin and Huxley. In panel A we see $n_\infty = (\bar{G}_0/\bar{g})$ vs $\bar{V}$, showing a sigmoidal increase in the number of open channels as $\bar{V}$ becomes more depolarized. Panel B illustrates $n'_\infty = (\bar{G}_1/\bar{g})$ which is the derivative of the curve in panel A. In panel C we see the normalized second derivative

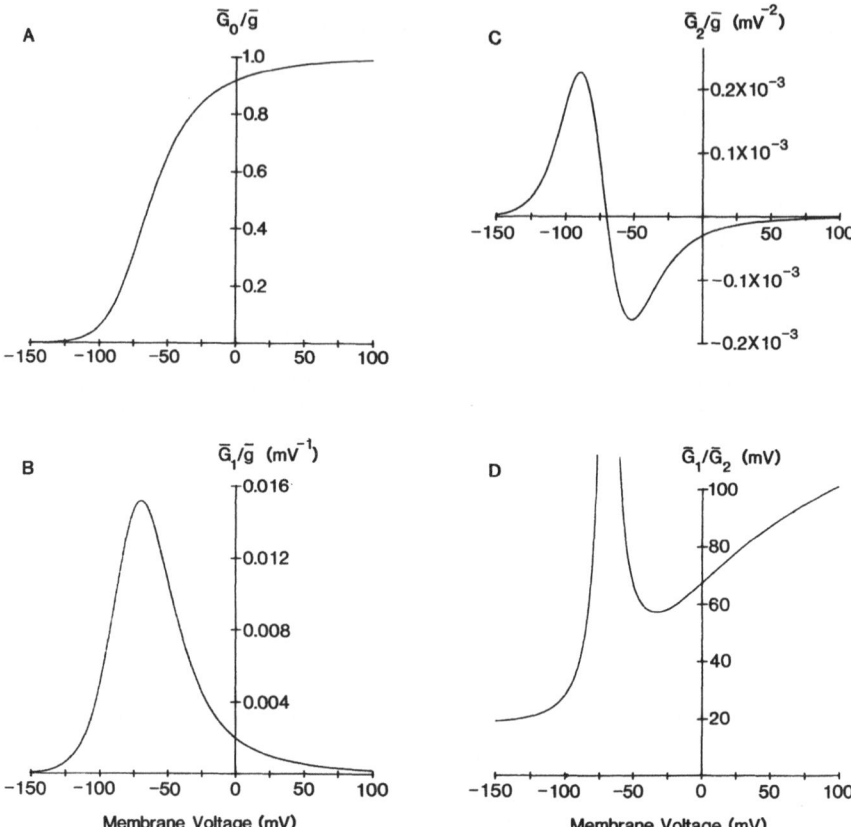

Fig. 6.   The voltage dependence of the weighting on the first
           three Volterra kernels describing the n process for the
           potassium channel in squid.   A: The steady state
           probability that the n gate will open as a function of
           the membrane voltage.   B: The voltage dependence of the
           scale factor for the kernel decribing the linearized
           response of the n process.   The abcissa is the steady
           state holding voltage whereas the ordinate is the
           scaling or weighting of the linearized response to small
           voltage perturbations about the steady state holding
           voltage.   C: The weighting of the second order (first

$\frac{1}{2}n_{\infty}'' = |\bar{G}_2/\bar{g}|$. Finally, in panel D we see the ratio $|\bar{G}_1|/|\bar{G}_2|$, given in millivolts, which specifies the allowable level of input voltage for a linear response. Over most values of holding potential, this ratio is greater than 50 mV, implying that an input signal of $|\epsilon v_m| \leq 5$ mV will keep the nonlinear contribution below 10% of the linear response. The parameter $\bar{G}_2$ goes through zero at $V \approx -12$ mV, and in this region of holding voltages we clearly do not need to worry about nonlinear contributions from terms proportional to $\bar{G}_2$. In fact, the maximal value of $\bar{G}_1$ must (because of the derivative relationship) correspond to the zero value of $\bar{G}_2$. It is therefore only at hyperpolarized holding voltages, where the K^+ channel has not significantly turned on, or at depolarized holding voltages, where the channel is driven into saturation, that $|\bar{G}_1|/|\bar{G}_2|$ becomes rather small. Of course, in these potential ranges the absolute values of $\bar{G}_1$ and $\bar{G}_2$ are individually insignificantly small, so studying the linear response of a channel, with a smooth sigmoidal change in the number of open channels vs holding voltage, does not require impractically small levels of driving voltage.

Linearized Analysis

If the input voltage fluctuations are kept within the amplitude limit just derived, then a linear analysis is

nonlinear term) kernel. D: The ratio of the linear to "nonlinear" steady state weighting functions. The response of the n process can be considered essentially linear when the voltage perturbations are much smaller than this ratio; if the perturbations are some % of the ratio then the nonlinear contamination is approximately the same % of the response.

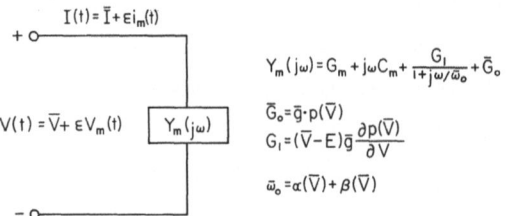

Fig. 7.   An illustration of the various linear components which
          comprise the membrane admittance. In this illustration,
          it is assumed that there is a single, first order,
          voltage and time dependent ionic channel.

appropriate and Figure 7 represents the linearized membrane
admittance. The membrane capacitance is assumed to be voltage
and frequency independent and is represented by C_m; the passive
leak conductance is called G_m and the membrane properties
conferred by the ensemble of first order channels are described
by the remaining two terms. The linearized admittance is defined
more formally by $I_1(j\omega)$ in equation (48):

$$Y_m(j\omega) \triangleq \frac{I_1(j\omega)}{V_m(j\omega)}$$

The term $\bar{G}_o$ in Figure 7 depends on the average conductance
contributed by the channels at the average membrane potential $\bar{V}$,
and so this term will most likely be absent at strongly
hyperpolarized membrane voltages, (see Fig. 6A). The frequency
dependent term is proportional to G_1, which is proportional to
several factors. This term arises because variations in the
average conductance of the channel allow variations in membrane
current that are driven by the steady electrochemical potential $\bar{V}$
- E. Hence this term is only important in the potential range

where $\bar{V}$ - E is significantly different from zero and in the range where the average number of open channels will change in response to voltage changes. For example, if all of the channels are open or if all of the channels are closed, then the conductance will not change in response to small voltage variations. Thus, for G_1 to be significantly different from zero, we must be in the potential range where $\partial p(\bar{V})/\partial v \neq 0$. The frequency dependent component of membrane conductance will therefore be absent at both strongly hyperpolarized and strongly depolarized membrane potentials. (see Fig. 6B).

The frequency dependence of the term proportional to G_1 arises because the channel conductance does not instantaneously respond to voltage changes, but has a characteristic time delay (given by $1/\bar{\omega}_0$). Because the channel is voltage dependent, the change in current lags behind the change in voltage and the linearized channel admittance resembles an inductance. If the current voltage relationship were current dependent, such as accumulation/depletion phenomena often associated with small extracellular compartments (Adrian and Freygang,1962; Barry and Adrian, 1973; Atwell et al., 1979; Almers et al., 1981), then the voltage change would lag behind the current change and the resulting linearized admittance would resemble a capacitance. The presence of a resonant peak in the membrane impedance can only occur when there are inductive effects, thus a voltage dependent channel has a characteristic imprint on the electrical impedance of a membrane or tissue.

It is interesting to compute the effects of a known channel (i.e. the Hodgkin and Huxley K^+ channel in squid) on the membrane admittance. These effects have been demonstrated experimentally by Cole (1972), and more recently by Fishman et al. (1982), so the simulations in Figure 8 are not just speculation.

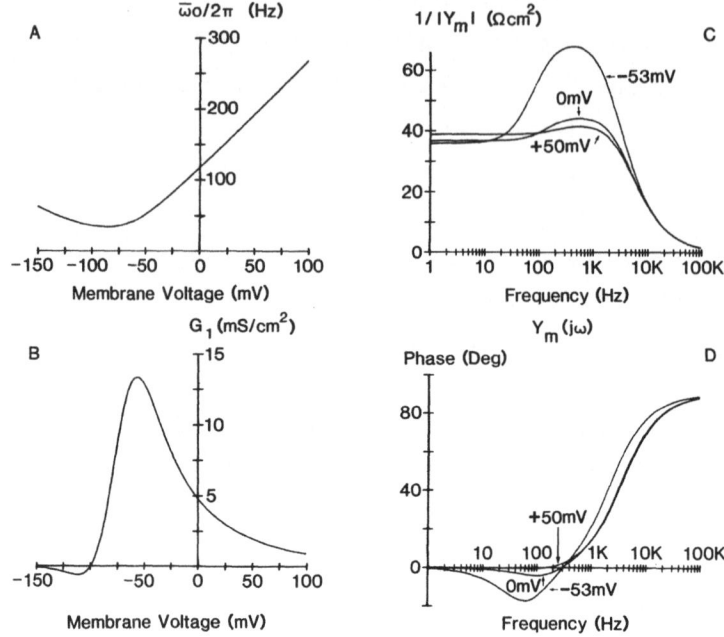

Fig. 8.   The impedance of the linearized membrane response
illustrated in Figure 7:   the impedance functions are
computed for three holding potentials (V = -53, 0, +50
mV).   The value of $\bar{\omega}_0$, G_1 can be determined from panels
A and B: the value of the average conductances of the
open channels, $\bar{G}_0$, can be determined by multiplying the
graph of $\bar{G}_0/\bar{g}$, given in Figure 6A, by $\bar{g}$ = 24 mS/cm^2 and
evaluating the result at the appropriate holding
voltage.   These curves represent the parameters and
impedance of a channel described by $i_k = (v-E_k)\bar{g}_n(v,t)$,
where E_k = -100 mV, and $n(v,t)$ is described by Hodgkin
and Huxley (1952) for the potassium channel in squid
axon.

The curves in Figure 8 are calculated using the parameters
of Hodgkin and Huxley (1952), for α_n and β_n, but with voltage in

the cell defined with respect to the bath, so their value of E_R = -58 mV and we assume E_K = -100 mV. In panel A we see the voltage dependence of $(\alpha_n + \beta_n)/2\pi$ which corresponds to $\bar{\omega}_0/2\pi$ in the theoretical analysis. And in panel B is the voltage dependence of the conductance (S/cm^2) contributed by fluctuations in the channel which are linearly related to the driving voltage. This parameter corresponds to G_1 in Figure 7. Notice the voltage dependence of G_1 will be skewed from that of $\bar{G}_1$ shown in Figure 6 because the current flowing through an ensemble of fluctuating channels will depend on the distance from the equilibrium voltage as well as the amplitude of the fluctuations.

In the Hodgkin-Huxley formalism, the probability function n is taken to the 4th power, hence g_K is not really a first order channel. However, the linearized response of any power function will be first order, as will be shown in the section on "Higher Order Channels." In the simulations in Figure 8 we have neglected the power law and assumed the conductance is directly proportional to n. This assumption hardly affects computations of the linear response, in fact it only changes the scale factor of G_1, but it does simplify computations of the Volterra correction term which is presented in the next section.

Panel C of Figure 8 illustrates the magnitude and phase of the linear membrane admittance shown in Figure 7. The resonant peak in the magnitude of the reciprocal admittance and the negative phase angle are characteristic of a membrane in which a voltage dependent channel is conducting current. The resonance is maximal at the voltage where the weighting conductance G_1 is maximal, and goes away at either more positive or more negative potentials.

The Volterra Correction Term

It is also useful to study the Volterra correction term for
a first order channel. First, the response $I_2(j\omega)$ will be
analytically presented and then $I_2(j\omega)$ is computed for a
Hodgkin-Huxley like K^+ channel. The computations will use the
parameters for a Hodgkin-Huxley K^+ channel, but the calculations
will be simplified by assuming $g_K = \bar{g}n$ rather than the more usual
n^4 formalism. It is noteworthy that the amplitude and <u>frequency</u>
dependence of the Volterra correction terms depend on the power
law, whereas only the amplitude of the linear response changes
with the power, (see the next section on "Higher Order
Channels").

The inherent difficulty in studying the nonlinear response
of a system resides in the relationship between the time course
of the input and the time course of the output. In general, one
cannot solve the differential equations that describe this
relationship. The results just presented however give a
systematic method of driving a differential equation, predicting
the response, and thereby identifying the parameters of the
equation. The method is to present a pure sinusoid of frequency
α_0 as an input, and observe the sinusoidal response at each
multiple $k\alpha_0$ (DeFelice et al., 1981). The response $I_2(j2\alpha_0)$
defines the dependence of the second harmonic on the parameters
of the differential equation. Similarly, higher order terms in
the Volterra series define the dependence of higher order
harmonics on the differential equation, (i.e., $I_k(jk\alpha_0)$ depends
only on $k\alpha_0$ and the parameters of the system). Difficulties
arise when a signal other than a pure sinusoid is presented to
the system. In this situation, equation (49) shows that I_2
depends on the integral over all combinations of input
frequencies taken two at a time. Similarly, I_k will depend on

the integral over all combinations of input frequencies taken k
at a time.

Consider first

$$v_m(t) = e^{j\alpha_0 t} \qquad (56)$$

In this situation $v_m(t)$ is a pure sinusoid of angular frequency
α_0 and peak amplitude of unity. The Fourier transform of
equation (56) is

$$V_m(j\omega) = \delta(\alpha_0 - \omega) \qquad (57)$$

where $\delta(\omega)$ is the Dirac delta function, which is probably most
meaningfully defined by its integral properties (Carrier et al.,
1966).

$$F(\alpha_0) = \frac{1}{2\pi} \int_{-\infty}^{\infty} F(\omega)\delta(\alpha_0 - \omega)d\omega \qquad (58)$$

$$\delta(\alpha_0 - \omega) = \int_{-\infty}^{\infty} e^{-j(\omega-\alpha_0)t} dt \qquad (59)$$

Notice that in equation (59) if $\omega \neq \alpha_0$ then we are integrating
around the unit circle an infinite number of times, but since
each circle contributes nothing to the integral, we get no
contribution for $\omega \neq \alpha_0$. However, if $\omega = \alpha_0$ then we integrate 1
over an infinite time and the integral is infinite. Thus, the
delta function is 0 for $\omega \neq \alpha_0$ and infinite at $\omega = \alpha_0$. This
behavior is worth noting since in practice we can only integrate
over a finite time T, thus our measured Fourier transform of a

sine wave will not be an infinite delta function, but will in
fact be a spike whose height is proportional to T and there will
be spillage over into other frequencies such that the area is
unity.

If the response $I_2(j\omega)$, given in equation (45), is computed
for an input sinusoid of frequency α_0, we obtain

$$I_2(j\omega) = \delta(2\alpha_0 - \omega)\left(\bar{G}_1 + \bar{G}_2[\bar{V} - E]\frac{1 + j\alpha_0/\chi\omega_0}{1 + j2\alpha_0/\bar{\omega}_0}\right)\frac{1}{1 + j\alpha_0/\bar{\omega}_0} \quad (60)$$

The response $I_2(j\omega)$ occurs only at $\omega = 2\alpha_0$, and by comparing
equation (60) with equation (51) one can see that (60) is related
to the second Volterra kernel $Y_2(j\omega_1,j\omega_2)$ by

$$I_2(j\omega) = \delta(2\alpha_0 - \omega)Y_2(j\alpha_0,j\alpha_0) \quad (61)$$

In the time domain, the observed second harmonic will be

$$i_2(t) = |Y_2(j\alpha_0,j\alpha_0)| \; e^{j(2\alpha_0 t + \phi_2(\alpha_0))} \quad (62)$$

where

$$\phi_2(\alpha_0) = \text{the phase angle of } Y_2(j\alpha_0,j\alpha_0).$$

Present technology (e.g. phase detectors or Fourier analyzers)
allows one to experimentally measure the amplitude and phase of
the second harmonic, thus if we apply sinusoids, <u>one at a time</u>,
of frequencies $k\alpha_0$: $k = 1, 2, \ldots, K$, then our measurement of
$Y_2(jk\alpha_0,jk\alpha_0)$ can be used to study the channel. Such data
provide information on the order of the kinetics of the channel
which cannot be obtained from the linear response.

Consider next an experiment in which one is only interested in the frequency dependence of the linear response. In this situation it is most expeditious to apply all of the sine waves simultaneously in the form of a pseudo random signal or true noise. Assume we apply a pseudo random signal of the form

$$v_m(t) = \sum_{k=1}^{K} e^{j(k\alpha_0 + \theta_k)} \tag{63}$$

where the phase angles θ_k are random numbers, $0 \le \theta_k \le 2\pi$. Equation (63) has equal weight at each frequency $\omega = k\alpha_0$, but because of the random phase relationship between the components, $v_m(t)$ will appear random within the time interval $0 \le t \le 2\pi/\alpha_0$, see Figure 1C.

The Fourier transform of equation (63) is

$$V_m(j\omega) = \sum_{k=1}^{k} \delta(k\alpha_0 - \omega)e^{j\theta_k} \tag{64}$$

The response $I_2(j\omega)$, defined in equation (45), to the input (64) is

$$I_2(j\omega) = \sum_{l=1}^{K} e^{j\theta_l} \sum_{k=1}^{K} \delta((1 - k)\alpha_0 - \omega)Y_2(jk\alpha_0, j l\alpha_0)e^{j\theta_k}$$

Thus, the second order response occurs at frequencies $2\alpha_0$, $3\alpha_0$, ..., $2K\alpha_0$. We can reorder the summation to give the weighting at each of these frequencies:

$$I_2(j\omega) = \sum_{1=2}^{2K} \delta(1\alpha_0 - \omega)y_2(j1\alpha_0) \tag{65}$$

where

$$y_2(j1\alpha_0) = \sum_{(k)} Y_2(jk\alpha_0, j(1 - k)\alpha_0)e^{j(\Theta_k + \Theta_{1-k})}$$

and $\sum\limits_{(k)}$ means summation over all values of k such that

$$1 \le k, \quad 1 - k \le K$$

The second order response to a pseudo random input of frequencies α_0, $2\alpha_0$, ..., $K\alpha_0$ is clearly quite different from the second harmonic recorded when pure sinusoids are applied, one by one, at the frequencies α_0, $2\alpha_0$, ..., $K\alpha_0$.

Comparing the results of these two possible experiments shows: (1) If a pure sinusoid is applied at the frequency α_0, then the response recorded at the input frequency α_0 is due to the linearized transfer function, the response recorded at $2\alpha_0$ is $Y_2(j\alpha_0, j\alpha_0)$ and the response at $k\alpha_0$ is $Y_k(j\alpha_0, ..., j\alpha_0)$. Thus, if we apply a pure sinusoid then we can record a response at all multiples of the basic harmonic and if one performs the tedious task of computing $Y_k(j\omega_1, ..., j\omega_k)$, then we can predict the response at each harmonic. (2) If we apply a pure sinusoid but vary the frequency so that we obtain a response due to inputs at α_0, $2\alpha_0$, ..., $K\alpha_0$, then the frequency dependence of each Volterra kernel can be studied. For example, if we stimulate at α_0 while recording at $2\alpha_0$, stimulate at $2\alpha_0$ while recording at $4\alpha_0$, ..., stimulate at $K\alpha_0$ while recording at $2K\alpha_0$, we can construct the frequency dependence of $Y_2(j\omega, j\omega)$. However, knowledge of

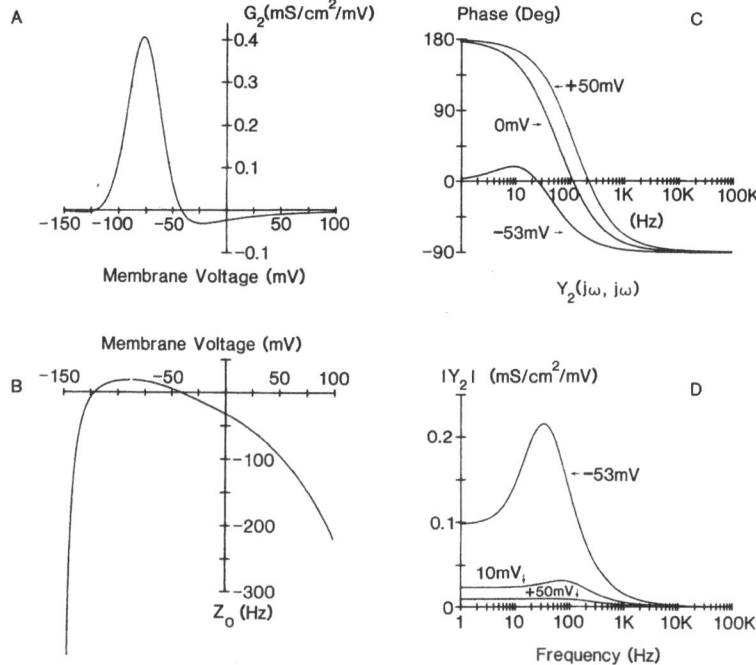

Fig. 9. The second order admittance $Y_2(j\omega, j\omega)$ measured by applying sine waves, one at a time, at a number of different frequencies: $k\alpha_0$, $k = 1, 2, \ldots, K$; and recording the second harmonic of the current response at frequencies $2k\alpha_0$ when the applied voltage sine wave is at $k\alpha_0$. The second order admittance is graphed for three different holding voltages (V = -53, 0, +50 mV) and the parameters that describe the admittance at each voltage are given by panels A and B of this figure and panel A of Fig. 8. Also see equations (66) of the text and the related discussion. Once again, these calculations are for a modified (n vs n^4) potassium channel in squid axon.

$Y_2(j\omega,j\omega)$ does not specify the response to a general input, such as a pseudo random signal, but it does allow us to study the differential equation which produces the nonlinearity. (3) If we apply a pseudo random input which has equal energy at the frequencies α_0, $2\alpha_0$, ..., $K\alpha_0$, then the response at α_0 will be the linearized response $I_1(j\alpha_0)$, the response at $2\alpha_0$ will be due to $I_1(j2\alpha_0) + \epsilon I_2(j2\alpha_0)$, and the response at $k\alpha_0$, $k \leq K$ will be the sum

$$\sum_{l=1}^{k} \epsilon^{l-1} I_1(jk\alpha_0).$$

Moreover, to further complicate matters, each response $I_1(jk\alpha_0)$ depends on all permutations of the input frequencies α_0, $2\alpha_0$, ..., $k\alpha_0$ taken 1 at a time, see equation (65). It therefore seems important to keep the amplitude of a pseudo random or random input sufficiently small so we can neglect all the nonlinear Volterra kernels.

Figure 9 illustrates the second order response of a Hodgkin-Huxley-like K^+ channel, when the sine waves are applied one at a time. This is the same first order channel that was used to generate Figure 8. The parameters illustrated in Figure 9 can all be experimentally measured (DeFelice et al., 1981) by observing the second harmonic of the current response, when a pure sine wave is applied as the driving voltage. The second order response given by equation (60) can be written in the following form:

$$Y_2(j\omega,j\omega) = \frac{G_2(1 + j\omega/z_0)}{(1 + j\omega2/\bar{\omega}_0)(1 + j\omega/\bar{\omega}_0)} \tag{66}$$

where

$$G_2 = \bar{G}_1 + \bar{G}_2 (\bar{V} - E)$$

and

$$z_0 = \chi\bar{\omega}_0 / \left(\frac{2\chi\bar{G}_1 + \bar{G}_2(\bar{V} - E)}{G_2} \right)$$

Panels A and B of Figure 9 illustrate respectively the voltage
dependence of G_2 and z_0; panels C and D show the magnitude and
phase angle of $Y_2(j\omega,j\omega)$, computed at holding voltages of
$\bar{V} = -53, 0, + 50$ mV.

Higher Order Channels

The analysis has thus far focused on the most simple
circumstance, where the only nonlinear current is due to a first
order channel. The results are somewhat more general than they
may seem however, because any channel that depends on multiple
but independent gates can be analyzed by the equations of the
preceding section. For example, consider a channel that requires
two independent reactions to occur before it is in the open
state. If the probability of each reaction having occurred is
respectively p and q, then the probability of a conducting state
is pq, and the conductance is given by

$$g(v,t) = \bar{g} \ p(v,t) \ q(v,t)$$

We may expand both $p(v,t)$ and $q(v,t)$ in a Volterra series, then
mulitply them together and collect terms of like powers in ε.
The probability of an open channel is

$$p(\bar{V} + \varepsilon v_m, t) \ q(\bar{V} + \varepsilon v_m, t) = [p_0 + \varepsilon p_1 + \varepsilon^2 p_2 + \ldots]$$

$$x \ [q_0 + \varepsilon q_1 + \varepsilon^2 q_2 + \ldots] \qquad (67)$$

and the linearized frequency domain response is

$$I_1(j\omega) = G_m + \bar{g} \ \bar{P}_0 \bar{Q}_0 + j\omega C_m + \bar{g} \ [\bar{V}-E] \left(\frac{\bar{Q}_0 \bar{P}_1}{1 + j\omega/\bar{\omega}_p} + \frac{\bar{P}_0 \bar{Q}_1}{1 + j\omega/\bar{\omega}_q} \right) (68)$$

The parameters above are defined analogously with equations (41) and (42).

If the rate constants for the reactions p and q, or for any number of independent series gates, are identical, such as the $n(v,t)$ processes in the Hodgkin-Huxley formalism for a K^+ channel, then the frequency dependent terms due to the channel will combine into a single term with only one time constant.

For example, if

$$g(v,t) = \bar{g} p^2(v,t) \qquad (69)$$

then

$$g(\bar{V} + \varepsilon v_m, t) = \bar{g}[p_0^2 + \varepsilon 2 p_0 p_1 + \varepsilon^2 (p_1^2 + 2 p_0 p_2) + \varepsilon^3 (2 p_3 p_0 + 2 p_1 p_2) + \ldots] \qquad (70)$$

If equation (70) for a p^2 channel is compared with equation (39) for a first order channel, we see that in each instance the first term of the series is a constant, the next (linear) term in each series has a single time constant but the third (first nonlinear Volterra correction) term differs in form in these two series. The third term depends only on p_2 in a first order channel whereas it depends on $p_1{}^2 + 2p_0p_2$ in a second order p^2 channel. Because the frequency dependency contributed by $p_1{}^2$ is different from that due to p_2, we can detect the second order behavior by studying the frequency dependence of $I_2(j\omega)$. We might next want to determine whether the channel is second order or third order. In a third order channel

$$g(v,t) = \bar{g}p^3(v,t)$$

and

$$g(\bar{V} + \epsilon v_m, t) = \bar{g}[p_0^3 + \epsilon 3p_0^2 p_1 + \epsilon^2(3p_0 p_1^2 + 3p_0^2 p_2) +$$

$$+ \epsilon^3(3p_0^2 p_3 + 6p_0 p p_2 + p_1^3) + \ldots] \quad (71)$$

Comparison of (71) with (70) shows that the lowest order difference in frequency dependence arises in the term proportional to ϵ^3, which contains $p_1{}^3$ in equation (71) but not in (70). Terms of lower order in ϵ differ by a scale factor, but the scale factor is sufficiently arbitrary that it alone may not be convincing proof of the order of the channel. One can carry on with higher power laws, but the result has already emerged from the analysis just presented: the Volterra series for a kth order power function differs significantly from that for a $(k - 1)^{st}$ order power function in the kth term.

Thus far, our modelling of channel gates has assumed that the probability of an open channel depends on independent conformational changes, and that if the changes in state are almost identical, then the small signal response will reduce to a single time constant. Many other models of channel gating have been proposed in order to describe the nonlinear behavior of membrane currents. For example, the opening of Na^+ channels has often been described by a temporally sequential set of reactions

(Armstrong and Bezanilla, 1977; Armstrong and Gilly, 1979;
Goldman and Hahin, 1978). In this type of model, time delays and
multiple time constants are obtained by delaying the number of
reactants available to the final reaction. This is done by
imposing the requirement that a gate must be in a particular
chemical state before it can open, a chemical state which is the
result of one or more previous reactions. An example of a second
order scheme is presented below.

$$
\begin{array}{ccc}
 & \overset{\alpha_q}{\rightarrow} & \overset{\alpha_p}{\rightarrow} \\
1 - p - q & q & p \\
 & \overset{\beta_q}{\leftarrow} & \overset{\beta_p}{\leftarrow}
\end{array}
\qquad\qquad g(v,t) = \bar{g}\, p(v,t)
$$

closed   closed   open

If one writes a 2nd order differential equation to describe
p(t), the coefficients of this equation will contain time
derivatives of the α's and β's. A first order differential
equation for the vector (p, q) seems to better illustrate the
physical processes.

$$
\frac{d}{dt} \begin{pmatrix} p \\ q \end{pmatrix} = \begin{pmatrix} -\beta_p & \alpha_p \\ \beta_p - \alpha_q & -(\alpha_p + \omega_q) \end{pmatrix} \begin{pmatrix} p - P_o \\ q - Q_o \end{pmatrix} \tag{72}
$$

and

$$
P_o(v) = \frac{\alpha_p \alpha_q}{\alpha_p \alpha_q + \beta_p \omega_q} \quad , \quad Q_o(v) = \frac{\alpha_q \beta_p}{\alpha_p \alpha_q + \beta_p \omega_q} \tag{73}
$$

$$
\omega_p = \alpha_p + \beta_p \quad , \quad \omega_q = \alpha_q + \beta_q
$$

The terms $P_o(v)$ and $Q_o(v)$ represent the steady state fraction of
reactants in each kinetic state. Define the vector quantities

$$\pi = \begin{pmatrix} p \\ q \end{pmatrix}; \quad \Omega_o = -\begin{pmatrix} -\beta_p & \alpha_p \\ \beta_p - \alpha_q & -(\alpha_p + \omega_q) \end{pmatrix}; \quad \Pi_o = \begin{pmatrix} P_o \\ Q_o \end{pmatrix} \tag{74}$$

Then

$$\frac{d}{dt}\pi = -\Omega[\pi - \Pi_o] \tag{75}$$

If we now consider the response to the voltage $\bar{V} + \epsilon v_m$, we can expand each of the vector quantities in (74). Viz.:

$$\Pi_o(v) = \bar{\Pi}_o + \bar{\Pi}_1 \epsilon v_m(t) + \bar{\Pi}_2 \epsilon^2 v_m^2(t) + \ldots \tag{76}$$

$$\bar{\Pi}_o = \Pi_o(\bar{V}); \quad \bar{\Pi}_1 = \frac{\partial}{\partial v}\Pi_o(\bar{V}); \quad \ldots$$

$$\Omega_o(v) = \bar{\Omega}_o + \bar{\Omega}_1 \epsilon v_m(t) + \bar{\Omega}_2 \epsilon^2 v_m^2(t) + \ldots \tag{77}$$

$$\bar{\Omega}_o = \Omega_o(\bar{V}); \quad \bar{\Omega} = \frac{\partial}{\partial v}\Omega_o(\bar{V}); \quad \ldots$$

$$\pi(v,t) = \bar{\Pi}_o + \epsilon\rho_1(t) + \epsilon^2\rho_2(t) + \ldots \tag{78}$$

Differentiation of a vector or matrix is simply differentiation of each component. The equation for the linearized or order ϵ response is:

$$\dot{\rho}_1 = -\bar{\Omega}_o[\rho_1 - \bar{\Pi}_1 v_m(t)] \tag{79}$$

Taking the Fourier transform gives

$$\rho_1^\dagger(j\omega) = [j\omega T + \bar{\Omega}_o]^{-1}\bar{\Omega}_o\bar{\Pi}_1 V_m(j\omega) \tag{80}$$

where

$$T = \begin{bmatrix} 1 & 0 \\ 0 & 1 \end{bmatrix}$$

$$[j\omega T + \bar{\Omega}_o]^{-1} = \frac{1}{(j\omega)^2 + (\bar{\omega}_p + \bar{\omega}_q)j\omega + \bar{\alpha}_p\bar{\alpha}_q + \bar{\beta}_p\bar{\omega}_q}$$

$$\times \begin{bmatrix} j\omega + \bar{\alpha}_p + \bar{\omega}_p & \bar{\alpha}_p \\ \bar{\beta}_p - \bar{\alpha}_q & j\omega + \bar{\beta}_p \end{bmatrix}$$

and the first order vector kernel is given by $\Pi_1(j\omega)$,

$$\Pi_1(jw) = [j\omega T + \bar{\Omega}_o]^{-1} \bar{\Omega}_o \bar{\Pi}_1 \tag{81}$$

It is now straightforward to carry out the matrix multiplications indicated in equation (81) and thereby obtain the linearized frequency domain expression for the probability the channel is open.

$$P_1(j\omega) = \frac{(\bar{P}_1\bar{\beta}_p - \bar{Q}_1\bar{\alpha}_p)j\omega + (\bar{\alpha}_p\bar{\alpha}_q + \bar{\beta}_p\bar{\omega}_q)\bar{P}_1}{(j\omega)^2 + (\bar{\omega}_p + \bar{\omega}_q)j\omega + \bar{\alpha}_p\bar{\alpha}_q + \bar{\beta}_p\bar{\omega}_q} \tag{82}$$

Although equation (82) has not been factored, it is clearly of the same form as the linearized response of two independent series gates given in (68). Moreover, substitution of the rate constants

$$\bar{\alpha}_q = 2\bar{\alpha}_p = 2\bar{\alpha}$$

$$2\bar{\beta}_q = \bar{\beta}_p = 2\bar{\beta} \tag{83}$$

reduces equation (82) to a single term with only one time
constant. This substitution was first noted by Bezanilla and
Armstrong (1975) for the Na$^+$ channel. Because such a
substitution is always possible, one cannot use linear admittance
data to distinguish multiple independent gates (in series) from a
single multi state gate, although independent gates in series
does seem more likely than the fortuitous relationship between
rate constants shown in (83).

The second order kernel for the two state model can be
derived from the differential equation for the vector
$\rho_2(t; \bar{V}, v_m)$. If we collect the order ε^2 terms when equations
(76), (77) and (78) are substituted into (75), then

$$\frac{d}{dt} \rho_2 = -\bar{\Omega}_o[\rho_2 - \bar{\Pi} v_m^2] - \bar{\Omega}_1[\rho_1 - \bar{\Pi}_1 v_m]v_m \tag{84}$$

Fourier transforming (84) gives

$$\rho_2^{\dagger}(j\omega) = [j\omega T + \bar{\Omega}_o]^{-1}[\bar{\Omega}_o\bar{\Pi}_2 + \bar{\Omega}_1\bar{\Pi}_1] \frac{1}{2\pi} \int_{-\infty}^{\infty} v_m(j\zeta)v_m(j\omega - j\zeta)d\zeta$$

$$- \bar{\Omega}_1 \frac{1}{2\pi} \int_{-\infty}^{\infty} \rho_1^{\dagger}(j\zeta)v_m(j\omega - j\zeta)d\zeta \tag{85}$$

Where the first order vector response $\rho_1^{\dagger}(j\omega)$ is defined in (80).
We can work backwards from (85) to define the second order vector
kernel $\Pi_2(j\omega_1, j\omega_2)$

$$\Pi_2(j\omega_1, j\omega_2) = \frac{1}{2}[j(\omega_1 + \omega_2)T + \bar{\Omega}_o]^{-1} \{2[\bar{\Omega}_o\bar{\Pi}_2 + \bar{\Omega}_1\bar{\Pi}_1]$$

$$- \bar{\Omega}_1[\Pi_1(j\omega_1) + \Pi_1(j\omega_2)]\} \tag{86}$$

where $\Pi_1(j\omega)$ is the first order vector kernel. It is now
straightforward but tedious to carry out the matrix
multiplications in (86) and obtain the second order kernel
$\rho_2^+(j\omega_1,j\omega_2)$ for the open state and membrane current. Moreover,
the vector approach is readily extended to situations where there
are more than two states.

Summary

The analysis of membrane impedance data requires a model of
membrane channels, and it is difficult to uniquely choose any
particular model using electrical measurements alone. The models
analyzed in this paper are archetypes of the most popular
schemes, but they obviously neglect many factors, such as ion-ion
interactions, convergence resistance at the channel opening,
electrostatic effects of membrane bound charges and so on (see
Andersen (1978) for a review of many of these effects). One
should keep these factors in mind when constructing a model of
channel gating, since they will almost certainly induce subtle
variations in channel kinetics. Although the addition of new
kinetic states in the gating scheme might describe some such
subtle variations, the resulting model will have no physical
basis and is therefore hardly more valuable than the data. The
kinetic schemes analyzed in this chapter represent crude, or
first approximations to channel behavior, yet they have been so
successful in the past that one suspects they have some real
physical basis.

Membrane channels are inherently nonlinear functions of
voltage, but like any nonlinear system, the degree of
nonlinearity depends upon the amplitude of the input. The
channels will behave linearly for very small inputs; if the
amplitude of the input is increased, the first observable
nonlinear effect will be proportional to the voltage squared and

so on. Because we are interested in small signal impedance
analysis, the Volterra series provides a suitable theoretical
framework to bridge between the nonlinear differential equations
describing the membrane current and the small signal impedance
data. The first term in the Volterra series is the linear
membrane impedance and the second term may be used to analyze
small nonlinear effects. The analysis of channel properties has,
in the past, been most often accomplished in the time domain on
the transient current following a step change in membrane
potential. Once the membrane potential has been clamped to some
steady voltage where the channels are activated, the rate
constants for the reactions opening the channels will remain
constant (i.e. rate constants are assumed to be instantaneous
functions of the membrane voltage). A small signal impedance
analysis, at the same steady voltage, will depend upon the same
constant values of the rate constants. Hence, one expects the
two experiments to yield comparable data. Nonetheless, the large
signal current recorded when the membrane potential is stepped
from a hyperpolarized value, where the channels are inactivated,
to a more depolarized value, where the channels are activated,
will have an initial slope proportional to the number of gates or
kinetic states, (e.g. if there are n gates, the initial slope of
$g(\bar{V},t)$ vs t is proportional to t^n), whereas a linearized
impedance analysis at the same steady depolarized voltage is
always first order, which implies a small step in voltage applied
on top of the holding potential will result in a conductance
change that has an initial slope proportional to t. The physical
basis of this difference, between the conductance changes induced
by small voltages and large voltage steps, is implied by models
of channel gating presented in this chapter.

The predicted difference does not occur because the step
response model and linearized model are intrinsically different;
rather, it is because the initial state of the membrane is

different in these two situations. Let us assume, for example,
that the linearized analysis is done at a potential where the
probability $P_o(\bar{V})$ of a gate being open is 50%. In this initial
state the p^2 membrane will have $P_o^2(\bar{V}) = 25\%$ of the channels with
both gates open, 50% of the channels with one gate closed, and
$(1 - P_o(\bar{V}))^2 = 25\%$ of the channels with both gates closed. When
we apply a small voltage increase, the resulting increase in the
number of open channels will come almost entirely from the pool
of channels with one gate open and one gate closed. Because this
is a first order transition, the resulting small signal linear
model is first order. If the membrane potential is initially at
a value where all of the p^2 channels have both gates closed, then
upon depolarization, the time course of the conductance will have
a second order delay, since two gates must open before any
channel can conduct current.

One important choice is which experimental protocol will, in
principle, most likely answer the questions one wants to address?
Clearly, the above discussion shows a linear impedance analysis
will not determine the number of gates within a channel, at least
not if the gates have similar kinetics. Moreover, the linearized
channel model interacts with other membrane parameters to produce
the measured impedance data, whereas step response transient data
are directly related to the channel's gating. Hence, if one has
an ideal situation, where a membrane can be voltage clamped
without series resistance or spatial inhomogeneities, then
voltage steps are the best choice of stimulus. However, even in
the ideal situation, it is worthwhile to experimentally verify
that models derived from large signal step analysis will indeed
predict the frequency domain response. And the Volterra series
offers a method of testing models that is quite different from
time domain measurements: the Volterra series relates the order
of the channel to the magnitude and phase shift of the harmonics
present in the channel's response to a pure sine wave.

One further restriction on the use of frequency domain techniques is that an impedance analysis can only be accomplished after the channel's transient behavior is complete, at least no one has yet defined a nonstationary impedance function similar to the nonstationary power spectrum analyzed by Sigworth (1981). Thus, if the channel turns on then off, such as the Na^+ channel, it may not be possible to obtain impedance data.

Despite the above mentioned restrictions, if one can obtain impedance data, then a wealth of information will be available concerning all of the geometrical factors and membrane parameters of the system. One of the real advantages of a linear analysis is that fast control of the membrane potential is not necessary. All of the channel parameters in the previous analysis are set by the d.c. potential $\bar{V}$, thus insofar as the d.c. potential is spatially uniform, the channels will be in a spatially uniform state. High frequency small membrane voltage perturbations can vary with location, such as down a tubule in skeletal muscle, or because of a series resistance in axons, but because the system is linear, we can account for series resistance or distributed resistances in our model. We can thereby quantitatively measure the series resistance and by doing a linear analysis, we have removed it as a potential source of artifact.

In addition to the ionic current which flows through a voltage dependent channel, the actual process of gating requires the movement of bound charges within the membrane. The gating current for the sodium channel in squid axon was first studied by Armstrong and Bezanilla (1973), using classical voltage clamp technology to observe the transient current which preceded or accompanied the opening of the channel. More recently, Fernandez et al. (1982) have used frequency domain techniques to characterize gating currents, and their results illustrate the

feasibility as well as the value of studying the linear impedance
function generated by gating currents.

Thus, although a linear impedance analysis should not
replace a step analysis, it does represent an independent
experimental technique for studying channel properties. If our
models of channel gating are correct, then we must be able to
explain frequency domain data or we must change our models.
Furthermore, the geometrical sources of artifact in a step
analysis can be quantitatively measured and often removed as
sources of artifact in a linear impedance study.

Tissues which contain small trapped extracellular spaces,
such as the T-system in skeletal muscle or clefts between cells
in heart muscle, are obvious candidates for a linear analysis.
In these tissues one cannot obtain fast, spatially uniform
control of the potential across inner membranes. Moreover, the
geometry of the tissue may vary considerably from preparation to
preparation and one needs the structural information that is
inherent in impedance data. The structural complexities present
in these tissues are likely to play an important role in their
overall electrical behavior, thus the impedance data may
highlight important interactions between membrane channels,
membrane capacitance and series resistance.

The techniques of voltage clamp, patch clamp, step analysis
and impedance analysis are complementary tools for studying
channels, membrane and tissues. We must be able to explain the
data recorded from ensembles of channels, in terms of the single
channel conductance, and we must explain the properties of a
tissue in terms of the specific properties of its membranes and
its geometry. Moreoever, the models of channel gating, which
have been derived from step analysis experiments, rigorously
dictate the behavior of the channel in the frequency domain, so

these models must conform to frequency domain experimental data, or they must be modified.

ACKNOWLEDGEMENTS

I would like to acknowledge the many contributions of Dr. R. S. Eisenberg to the development and understanding of impedance measurements using microelectrodes.

This work was supported in part by grants HL 29205 and EY03095 from the National Institutes of Health and American Heart Association grant 79-851 with funds contributed in part by the Chicago Heart Association.

REFERENCES

Abramowitz, M. and Stegun, I. A., 1972, "Handbook of Mathematical Functions," Dover, N.Y.

Adrian, R. H. and Freygang, W. H., 1962, The potassium and chloride conductance of frog muscle membrane, J. Physiol., 163:61.

Almers, W., Fink, R. and Palade, P.T., 1981, Calcium depletion in frog muscle tubules: the decline of calcium current under maintained depolarization, J. Physiol., 312:177.

Andersen, O. S., 1978, Permeability properties of unmodified lipid bilayer membranes, in: "Membrane Transport in Biology, Vol. I, "D. C. Tosteson, ed., Springer-Verlag, Berlin.

Armstrong, C. M. and Bezanilla, F., 1973, Currents related to movement of the gating particles of the sodium channels, Nature, 242:459.

Armstrong, C. M. and Bezanilla, F., 1977, Inactivation of the sodium channel. II. Gating current experiments. J. Gen. Physiol., 70:567.

Armstrong, C. M. and Gilly, W. F., 1979, Fast and slow steps in the activation of sodium channels, J. Gen. Physiol., 74:691.

Attwell, D., Eisner, D. and Cohen, I., 1979, Voltage clamp and tracer flux data: effects of a restricted extra-cellular space, Quart. Rev. Biophys., 12:213.

Barry, P. H. and Adrian, R. H., 1973, Slow conductance changes due to potassium depletion in the transverse tubules of frog muscle fibers during hyperpolarizing pulses, J. Membrane Biol., 14:243.

Bendat, J. S. and Piersol, A. G., 1971, "Random Data: Analysis and Measurement Procedures," Wiley, N.Y.

Bezanilla, F. and Armstrong, C. M., 1975, Kinetic properties of the gating currents of sodium channels in squid axon, Phil. Trans. R. soc. Lond. B., 270:449.

Carrier, G. F., Krook, M. and Pearson, C. E., 1966, "Functions of a Complex Variable," McGraw-Hill, N.Y.

Clausen, C. and Fernandez, J. M., 1981, A low-cost method for rapid transfer function measurements with direct application to biological impedance analysis, Pflugers Arch., 390:290.

Cole, K. S., 1972, "Membranes Ions and Impulses," University of California Press, Berkeley.

DeFelice, L. J., Adelman, Jr., W. J., Clapham, D. E. and Mauro, A., 1981, Second-order admittance in squid axon, in: "The Biophysical Approach to Excitable Systems," W. J. Adelman, Jr., ed., Plenum, N.Y.

Desoer, C. A. and Kuh, E. S., 1969, "Basic Circuit Theory," McGraw-Hill, N.Y.

Eisenberg, R. S. and Mathias, R. T., 1980, Structural analysis of electrical properties of cells and tissues, in: "Critical Reviews in Bioengineering," J. R. Bourne, ed., CRC Press, Boca Raton, FL.

Eisenberg, R. S., 1983, Chapter II. Impedance measurement of the
    electrical structure of skeletal muscle, in: "Handbook of
    General Physiology," L. D. Peachey, ed., Williams and
    Wilkens, Inc., Baltimore, MD.

Falk, G. and Fatt, P., 1964, Linear electrical properties of
    striated muscle fibers observed with intracellular
    electrodes, Proc. R. Soc. Lond. B. Biol. Sci., 160:69.

Fernandez, J. M., Bezanilla, F. and Taylor, R. E., 1982,
    Distribution and kinetics of membrane dielectric
    polarization, J. Gen. Physiol., 79:41.

Fishman, H. M., Moore, L. E. and Poussart, D., 1981, Squid axon K
    conduction: admittance and noise during short- versus
    long-duration step clamps, in: "The Biophysical Approach to
    Excitable Systems," W. J. Adelman, Jr., ed., Plenum, N.Y.

Goldman, L. and Hahin, R., 1978, Initial conditions and the
    kinetics of the sodium conductance in Myxicola giant axons,
    J. Gen. Physiol., 72:879.

Hodgkin, A. L. and Huxley, A. F., 1952, A quantitative
    description of membrane current and its application to
    conduction and excitation in nerve, J. Physiol., 117:500.

Kootsey, J. M. and Johnson, E. A., 1973, Buffer amplifier with
    femtofarad input capacity using operational amplifiers, IEEE
    Tran. Biomed. Engr., 20:389.

Levis, R., Mathias, R. T. and Eisenberg, R. S., 1983. Electrical
    properties of sheep Purkinje strands. Electrical and
    chemical potentials in the clefts. Biophys. J., 44:225.

Marmarelis, P. Z. and Naka, K. I., 1974, Identification of
    multi-input biological systems, IEEE Trans. on Biomed.
    Engr., 21:88.

Mathias, R. T., Rae, J. L. and Eisenberg, R. S., 1979, Electrical
    properties of structural components of the crystalline lens,
    Biophys. J., 25:181.

Mathias, R. T., Rae, J. L. and Eisenberg, R. S., 1981, The lens
    as a nonuniform spherical syncytium, Biophys. J., 34:61.

Mathias, R. T., Ebihara, L., Lieberman, M. and Johnson, E. A.,
    1981, Linear electrical properties of passive and active
    currents in spherical heart cell clusters, Biophys. J.,
    36:221.

Palm, G. and Poggio, T., 1977a, The Volterra representation and
    the Wiener expansion: validity and pitfalls, SIAM J.,
    33:195.

Palm, G. and Poggio, T., 1977b, Wiener-like system identification
    in physiology, J. Math. Biol., 4:375.

Papoulis, A., 1965, "Probability, Random Variables, and
    Stochastic Processes," McGraw-Hill, N.Y.

Papoulis, A., 1977, "Signal Analysis," McGraw-Hill, N.Y.

Poussart, D. J. M. and Ganguly, U. S., 1977, Rapid measurement of
    system kinetics - an instrument for real-time transfer
    function analysis, Proc. IEEE, 65:741.

Schetzen, M., 1980, "The Volterra and Wiener Theories of
    Nonlinear Systems," Wiley, N.Y.

Sigworth, F. J. 1981, Interpreting power spectra from
    nonstationary membrane current fluctuations, Biophys. J.,
    35:289.

Smith, J. I., 1971, "Modern Operational Circuit Design,"
    Wiley, N.Y.

Valdiosera, R., Clausen, C. and Eisenberg, R. S., 1974,
    Measurement of the impedance of frog skeletal muscle fibers,
    Biophys. J., 14:295.

Weiner, N., 1958, "Nonlinear Problems in Random Theory," Wiley,
    N.Y.

Zadeh, L. A., and Desoer, C. A., 1963, "Linear System Theory:
    The State Space Approach," McGraw-Hill, N.Y.

"STATIONARY" FLUCTUATIONS OF Na CURRENT IN MYELINATED NERVE

Wolfgang Nonner

Department of Physiology and Biophysics
University of Miami, P.O. Box 016430
Miami, FL 33101, USA

Current in Na channels of nerve or muscle is transient. Its activation upon a depolarizing stimulus is followed by a spontaneous inactivation. Study of Na current fluctuations with methods that require stationarity is nevertheless possible: inactivation tends to be incomplete, so that a measurable Na current persists over hundreds of milliseconds, long enough for recordings in a nearly stationary state. While this sustained current is small and unimportant for the propagation of the nervous impulse, it allows us to analyze gating at just those membrane voltages where the Na channels are most sensitive to the electric field.

Most of the published work on ensemble fluctuations from Na channels has been done on myelinated nerve of the frog. The first part of this paper discusses technical aspects of the "stationary" approach to this preparation as used by Conti et al. (1976, 1981). The second part discusses the spectral composition of stationary fluctuations with regard to kinetic models of Na channel gating.

TECHNIQUES

The Na channels of myelinated nerve tend to become
inactivated to a fairly large extent, so that any stationary Na
current is on the order of a hundredth of the maximal transient
current. For detecting this current, one needs to block the
irrelevant sustained current in K channels, as can readily be done
with intracellular Cs and extracellular tetraethylammonium ions.
The sustained Na current is then still overlaid by "leakage"
current which we do not know how to block. The leakage and
sustained Na currents must be separated then by criteria like
sensitivity to block by tetrodotoxin (TTX) or non-linear
dependence on voltage.

Subtracting records without and with TTX gives results
indistinguishable from those of a "P/2" procedure (Bezanilla &
Armstrong, 1977) for measuring and subtracting linear membrane
current. (In the form used here, the "P/2" protocol includes,
between groups of test sweeps with depolarizing pulses, control
sweeps with hyperpolarizing pulses that are of one half the
amplitude of test pulses. Twice the control current is
algebraically added to the test current.) The stationary Na
current is on the order of 200 pA or less. As a control, the
"P/2" procedure applied to a TTX-poisoned fiber yields at most a
few pA of net current. The "P/2" method has been preferred in
recent work as it is less sensitive than the TTX method to slow
changes in the leak conductance, which are often seen over the
long periods needed for noise measurements. It has, however, the
potential drawback that it must fail if substantial Na current is
present at the voltage from which the "P/2" pulses are applied.
At the resting potential, for instance, TTX eliminates about 20 pA
of Na inward current. When applied from this level, the "P/2"
method will estimate a net inward current that is up to 60 pA less
than the actual Na current during the depolarizing pulse. The

work of Neumcke et al. (1979), suggesting that the conductance in individual Na channels decreases as the holding potential is moved into the hyperpolarized region, may have been biased in this way.

The fluctuations of the small stationary Na current are relatively strong. If channels randomly and independently gate an individual current of 1 pA, the binomial theorem predicts a fluctuation of 10 pA r.m.s. about a mean of 100 pA. On the other hand, this fluctuation may be obscured by background noise from other sources.

In voltage-clamp setups for myelinated nerve, thermal noise arises from a variety of sources located in the fiber, nerve chamber, and feedback amplifier (for a detailed analysis see Conti et al., 1976). Selecting thick (>15 μm) axons, cutting the ends on both sides of the node to appropriate lengths, and rejecting nodes that give a large slow capacity current (Dodge & Frankenhaeuser, 1959) help keep the background noise low. Other useful improvements include an air gap, which replaces a critical vaseline seal in the chamber, and an amplifier that has low voltage noise and a small input capacitance. In a good experiment, the spectral density of the background noise can be kept near 10^{-26} A^2/Hz over 0 to 5 kHz, which amounts to an r.m.s. value of about 7 pA.

With 7 pA of background noise, a background correction of the measurements is necessary. To this end, power spectral densities, first measured while Na channels are operating and, again, after channels have been blocked by TTX, may be subtracted from each other, and their difference may be taken to represent noise in Na channels. It should be noted that this method requires that the membrane impedances be closely similar between the two sets of measurements, i.e. that the Na conductance be small with regard to leak conductance. Varying the total membrane impedance

appreciably would vary the size of the background noise in the
current signal (Conti et al., 1976), so the above subtraction
method is not applicable, for instance, to measurements on
transient Na current.

Na channels that are held depolarized over seconds or minutes
tend to become unavailable for opening by a slow process of
inactivation (Adelman and Palti, 1969; Peganov et al., 1973),
distinct from the fast inactivation described by Hodgkin and
Huxley (1952). If one wishes to study the gating processes that
give rise to the nervous impulse with little interference from the
slow inactivation, one needs to find a compromise for the length
of depolarization episodes. The minimal length includes a
settling time for the fast gating and one period of the lowest
fluctuation frequency of interest. A maximal length is reached
when Na current fades away by slow inactivation or when the slow
inactivation itself begins to contribute appreciable fluctuation.
Experiments in myelinated nerve have thus mostly been done with
pulse depolarizations lasting a fraction of a second and
alternating with recovery intervals of a few seconds. The method
prevents slow inactivation from accumulating, but necessarily
measures Na current that is non-stationary in its slow
inactivation. Indeed, the Na current and its fluctuation variance
often decrease appreciably over the pulse length needed for a
single spectrum. This complicates the analysis in two ways.

Firstly, the stochastic variations of the Na current need to
be separated from the systematic drift of the mean. With the
pulse technique, many similar records can be obtained. The time
course of the mean may be calculated by averaging a number of
subsequent records and may be subtracted from each individual
record. The mean may also be implicitely eliminated by
subtracting the first from the second of each pair of records.
The latter method is less sensitive to systematic variations of

the mean over many trials, but requires nearly twice as many
records for determining noise power spectral densities within
given limits of random error.

The second difficulty arises from the non-stationarity in the
stochastic part of the signal. If one proceeds in the usual way,
by Fourier-transforming the samples and taking the squares of the
moduli, one ends up with a particular kind of power spectrum that
in general cannot simply be compared to theoretical curves derived
for the stationary condition. For deriving stationary spectral
density functions, one usually writes down the stationary
autocovariance function and then takes the Fourier transform. For
deriving non-stationary spectral density, however, one has to
write down the non-stationary (i.e. two-dimensional)
autocovariance function, time-average the function over a period
corresponding to one experimental record, and then take the
Fourier transform (Conti et al. 1980; Sigworth, 1980). The
non-stationary autocovariance function used for slow-inactivating
Na channels needs to account for the classical gating processes as
well as for the slow inactivation.

Fortunately, both kinds of theoretical spectral curves can be
virtually indistinguishable from each other if certain conditions
are met, so experimental spectra can still be interpreted in terms
of the simpler, stationary theory. This is illustrated in Fig. 1
for a Na channel whose open probability is given by m^3hk, where m
and h are the first-order variables of the Hodgkin-Huxley
formalism and k is an analogous variable describing the slow
inactivation. Panel A shows the time courses of the mean current
for the two kinds of calculations done. These records are assumed
to start well after the fast gating (m^3h) has become stationary.
One record reveals a marked slow inactivation (triangles), the
other one does not (line). The time-averages of the mean currents
have been made identical by choosing appropriate numbers of

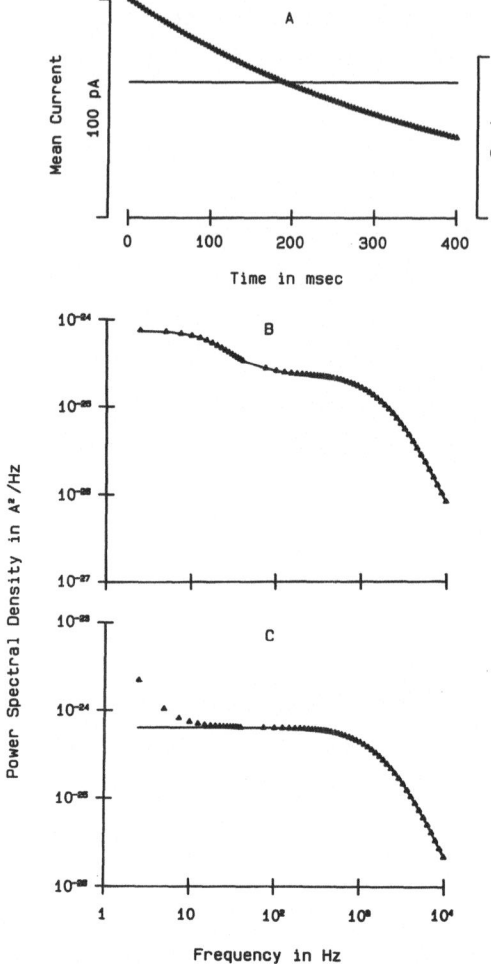

Fig. 1.   Comparison of stationary and non-stationary theories. Two
models are considered.   One model (solid lines) follows
Hodgkin-Huxley kinetics and is stationary.   The other
model (triangles) also has Hodgkin-Huxley kinetics and,
in addition, undergoes an independent, slow inactivation.
The latter process is non-stationary over the period of
interest.   (A) Mean current, computed as $Nim^3_\infty h_\infty$ (N =
62.231, i = 1 pA), or $Nim^3_\infty h_\infty k(t)$ ($N = 10^5$, i = 1 pA, k =
$k_\infty + (k_o - k_\infty) \exp(-t/\tau_k)$ where $k_\infty = 0$, $k_o = 1$, $\tau_k = 0.4$

channels for each ensemble. The theoretical spectral curve (Panel B) for the channels without slow inactivation (solid line) is indistinguishable from that for channels with slow inactivation (triangles). This result may be surprising, but it has been shown to apply at least to all kinds of channel models that include a kinetically unique open state and maintain a low open probability during the period of measurement (Sigworth, 1981). Panel C illustrates a case where the stationary theory is no longer applicable. The spectral curves again apply to channels without (line) and with (triangles) non-stationary slow inactivation, but the open probability has been sharply increased by eliminating the fast inactivation process (h = 1). Slow inactivation is now clearly reflected by an increased spectral density at the lower frequencies, and thus needs to be explicitly considered when theory is compared to experiment.

Records of "stationary" Na current obtained with the techniques described before may thus be treated as if they were perfectly stationary, even if a slow inactivation is evident. Caution is required, however, after channels have been treated in some way to block their fast inactivation process.

sec). Ensemble sizes, N, have been chosen to give identical time averages of the mean currents. Theoretical spectra have been calculated for the normal channels (B) and channels that have been modified to lack fast inactivation (C). Open probabilities $m_\infty^3 h_\infty$ are 0.001 in (B), and 0.027 in (C). The Hodgkin-Huxley parameters are the averages, for 24 mV depolarization, from Table 1 in Conti et al. (1980): $m_\infty = 0.3$, $h_\infty = 0.037$ (B) or $h_\infty = 1$ (C), $\tau_m = 0.17$ msec, $\tau_n = 7.3$ msec.

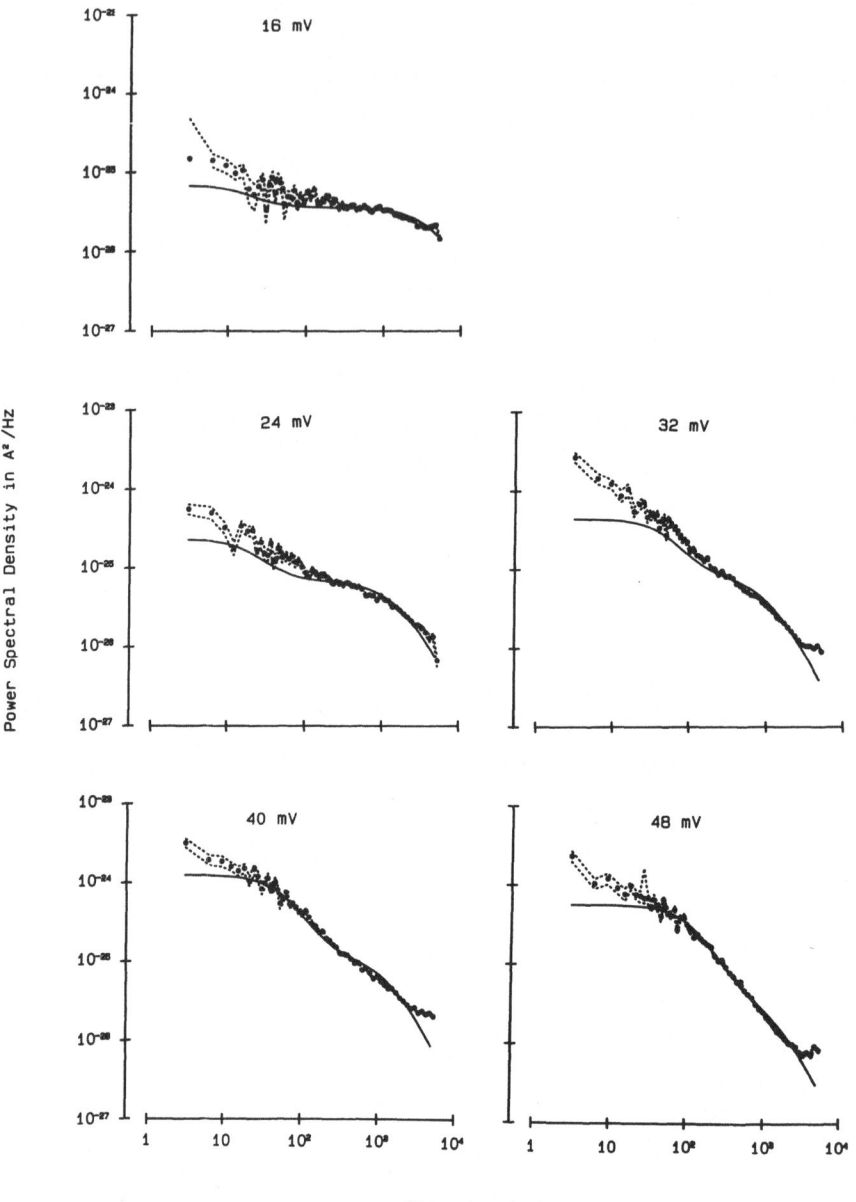

Fig. 2.   Na current fluctuation spectra measured at various
          depolarizations. Experimental power spectral densities
          (symbols) from Conti et al. (1980, their Fig. 3). Dotted
          lines marks ±1 SEM. The solid lines represent

EXPERIMENTAL POWER SPECTRAL DENSITY

Figure 2 gives examples of power spectra taken from the work
of Conti et al. (1980). The symbols represent the spectral
density attributable to fluctuations of "stationary" Na current in
the frog node of Ranvier. Dashed lines mark the range of ±1 SEM.
The spectral shapes are complex, like one expected from the gating
kinetics of Na channels, and vary considerably among the different
membrane voltages. At the smaller depolarizations, power is
spread rather evenly over the 3 Hz to 5 kHz range, whereas at the
larger depolarizations it tends to become localized in the low
frequencies.

The spectra span more than three decades of frequency.
Nevertheless, those from the smaller depolarizations give a fairly
incomplete view of the roll-off at high frequencies, suggesting
that the technique is missing an appreciable contribution of fast
fluctuations. Integration of the spectra over the accessible
frequencies is thus likely to underestimate the fluctuation
variance of the gating process. Dividing the mean current into
the experimental variance has yielded single-channel currents
corresponding to conductances of 3.8, 4.8, 7.4, 8, and 12.7 pS for
the 16 through 48 mV experiments in Fig. 2. On the assumption
that the actual conductance is constant and close to the upper
limit, one suspects that the part of the spectra not visible in

predictions of a model with Hodgkin-Huxley kinetics. The
parameters m_{∞}, h_{∞}, τ_m, τ_h in the curves were constrained
to the values found to describe the kinetics of transient
Na current in the same fibers; they were taken from Table
1 of Conti et al. Only amplitudes were adjusted by a
least-squares fit.

<u>Fig. 2 accounts for up to two thirds of the total fluctuation</u>
<u>variance.</u> The variance of ensemble fluctuations during the
transient Na current, also measured within the 5 kHz bandwidth,
reveals the same trend (Sigworth, 1980). Until higher frequencies
have been explored, interpretations should probably not be based
on the empirical variances measured at small depolarizations.
With regard to single-channel recordings, one may suspect that
published records, limited to frequencies no higher than 1 kHz,
have not yet exhibited much detail of the fast components of Na
channel gating.

The power spectra in Fig. 2, on the other hand, reveal
sufficient detail in their variation from voltage to voltage to
encourage tests of kinetic models. The limitations at high
frequencies need to be kept in mind, but some theoretical
inferences can be based on the known power spectral densities.

A MICROSCOPIC MODEL WITH HODGKIN-HUXLEY KINETICS

The solid lines in Fig. 2 have been predicted from a
microscopic version of the Hodgkin-Huxley (1952) description of Na
current. The channels are assumed to gate in the all-or-none
manner and are given the open probability m^3h. The shape of the
curves has been calculated from the parameters m_∞, h_∞, τ_m, and τ_h
found from fits of macroscopic, transient Na current in each fiber
(Table 1 of Conti et al., 1981). Only the vertical position
(involving knowledge of single-channel current or ensemble size)
has been chosen freely to give a least-squares fit to each set of
experimental points, taking into account the empirical standard
errors.

Two aspects are well predicted from this tentative model. The corner frequencies of the prominent Lorentzian components are in approximately correct positions, and the variation with voltage of the balance between the high- and low-frequency components is in principle reproduced. Nevertheless, the theoretical curves deviate in significant detail from the points. In particular, the theory tends to give an inadequate balance between the high- and low-frequency parts of the spectra.

This method of testing a model correlates two quite different kinds of measurements. Experimental spectra are taken from a small, non-inactivating fraction of Na current that is present, e.g., between 150 and 450 msec after the onset of a depolarization. Theoretical spectra are predicted from the kinetics of a large, transient Na current that is present within the first few milliseconds of a depolarizing step. A valid comparison can only be made if the individual Na channel is independent of its neighbors and if its gating mechanism does not become altered over a prolonged depolarization. The test also implies that one is dealing with a single ensemble of channels. The strongest experimental support for the first point comes from a recent experiment by Patlak and Horn (1982): they isolated several, closely spaced Na channels under a patch electrode and found the binomial distribution of current amplitudes expected for mutually independent channels. There is no direct experimental support to the second point, as the macroscopic kinetics of Na current before and after a depolarization of several hundred milliseconds still need to be compared. The single-channel conductances prevailing under either condition have, however, been verified to be indistinguishable (Conti et al., 1976; Sigworth, 1980). Plots of ensemble variance versus transient Na current are consistent with a single ensemble of channels (Sigworth, 1980).

FAST AND SLOW FLUCTUATIONS

The spectra in Fig. 2 indicate that Na current fluctuations reflect the kinetics of both major gating processes, activation and inactivation. Fast fluctuations attributable to the activation appear to be dominant at the small depolarizations, whereas slower fluctuations attributable to the inactivation seem to prevail at the larger depolarizations. The model with Hodgkin-Huxley kinetics predicts this tendency, so it may be useful in understanding the effect.

Fig. 3 shows computed simulations of how an individual Na channel with Hodgkin-Huxley kinetics may respond to depolarizations by 16 or 48 mV. In each case, the channel was initially in the resting state and then was allowed to move stochastically for 400 ms among all of its eight possible states. The top trace of each column indicates the inactivation status, and the second trace, the actual opening and closing. At 16 mV, the channel spends more time in non-inactivated states than at 48 mV, as is expected from the voltage dependence of inactivation. On the other hand, because of the opposite voltage dependence of the activation, actual openings at 16 mV are rare and short; they occur only in four of the non-inactivated periods. The third and following traces in Fig. 3 show these periods at a 50 times higher time resolution. At 48 mV, openings are likely to occur in each non-inactivated period and then form a burst. If we were to pass the single-channel signals through a low-pass filter whose time constant is similar to the length of a non-inactivated period, the filter would nearly abolish the single current pulses at 16 mV, whereas it would smooth out the bursts at 48 mV into current pulses that have the length of non-inactivated periods and an amplitude intermediate between zero and the full single-channel current.

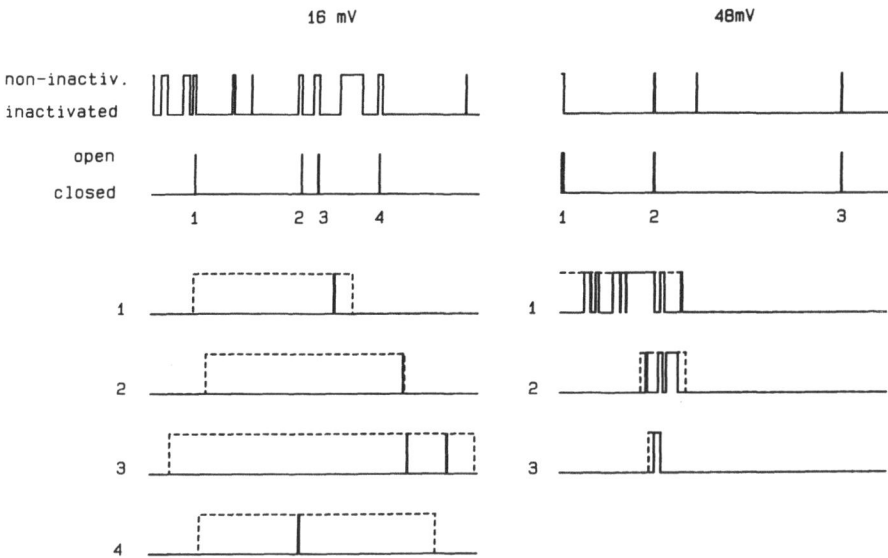

Fig. 3.   Simulation of single-channel activity from Hodgkin-Huxley
           kinetics. A non-inactivated, non-activated channel is
           exposed to a step depolarization of 16 or 48 mV; the
           resulting transitions are shown for a 400 msec period
           (top traces). The numbered traces represent 8 msec
           segments centered around times when the channel opens; in
           these traces non-inactivated periods are marked by the
           dashed lines, openings by the solid lines. The
           simulation was done by having a generator of random
           numbers choose every 5 μsec among the transitions that
           were possible among the eight states of a kinetic scheme
           that gave m^3h Hodgkin-Huxley kinetics. Rate constants
           were taken from Conti et al. (1980) and were the same as
           those for the corresponding spectral curves in Fig. 2.

       The low-pass filtered events are indistinguishable from those
of a channel that opens and closes with the characteristics of the
inactivation gating to pass a current equivalent to the average

current during the bursts seen in Fig. 3. They may be attributed
to the variance

$$\text{var}_{\text{slow}} = p_{\infty}(\text{not inactiv.})[1 - p_{\infty}(\text{not inactiv.})] \times$$
$$[p(\text{open/not inactiv.})i]^2 \qquad (1)$$

where $p_{\infty}(\text{not inactiv.})$ is an absolute probability, and $\bar{p}(\text{open/not}$
$\text{inactiv.})$ is the conditional open probability of the
non-inactivated channel, averaged over the non-inactivated period.
In the Hodgkin-Huxley model, the latter probability equals m^3_{∞},
and p_{∞} (not inactiv.) equals h_{∞}. Thus

$$\text{var}_{\text{slow}} = h_{\infty}(1 - h_{\infty})(m^3_{\infty}i)^2 \qquad (2)$$

In models in which the activation and inactivation processes are
kinetically interdependent, the conditional open probability will
generally vary during a non-inactivated period. Only if
rearrangements among activation states settle within a time that
is short relative to the non-inactivated periods, the time average
of the conditional open probability may be approximately
represented by the stationary value. The total variance of the
slow and fast fluctuations, on the other hand, of any open-close
mechanism is

$$\text{var} = p_{\infty}(\text{open})[1 - p_{\infty}(\text{open})]i^2, \qquad (3)$$

or in terms of the Hodgkin-Huxley model,

$$\text{var} = m^3_{\infty}h_{\infty}(1 - m^3_{\infty}h_{\infty})i^2 \qquad (4)$$

For the Hodgkin-Huxley model, the relative variance of the slow
fluctuations is then

$$r_{slow} = \frac{var_{slow}}{var} = \frac{(1 - h_\infty)m^3_\infty}{1 - m^3_\infty h_\infty} \tag{5}$$

and in the limit of small open probability, $m^3_\infty h_\infty \ll 1$,

$$r_{slow} = (1 - h_\infty)m^3_\infty \tag{6}$$

Since both factors, $1 - h_\infty$ and m^3_∞, increase with the depolarization, the relative variance of the slow fluctuations does so even more. This effect obviously accounts for much of the variation seen among the fitted curves in Fig. 2.

Combining eqns. (1) and (3) and using the identity

$$p_\infty(\text{open}) = p_\infty(\text{not inactiv.})\ \bar{p}(\text{open/not inactiv.}) \tag{7}$$

yields the analogous relative variance in a more general form:

$$r_{slow} = [1 - p_\infty(\text{not inactiv.})]\ \bar{p}(\text{open/not inactiv.}) \tag{8}$$

This equation may be used in considering models in which the activation and inactivation gating do not occur independently of each other.

The fluctuation spectra predicted from the Hodgkin-Huxley kinetics, though in general quite adequate, seem to be biased in underestimating the intensity of slow fluctuations (Fig. 2). The bias is considerable: for balancing the high- and low-frequency components more appropriately, the low-frequency Lorentzian term needs to be given up to four times more power than is consistent with the model (Conti et al., 1980). One may thus ask whether different kinetic models can be expected to yield more suitable values of r_{slow}, and whether in particular a kinetic coupling between the activation and the inactivation gating is important in determining that ratio.

Of the two terms that govern r_{slow} in eqn. (8), the first term is unlikely to vary much among reasonable models. It represents the steady-state inactivation and should be close to unity for the conditions under which the stationary fluctuations are measured. The second term represents open probability averaged between a time when the channel returns from inactivation and a time when it inactivates again. If a model yields a markedly larger r_{slow} than the Hodgkin-Huxley model, it needs to do so through the latter probability. There is, however, a further constraint: the new model should give correct open probabilities during a transient response as well. These probabilities are usually estimated by normalizing permeabilities with respect to a limiting permeability measured with a large depolarization. Sigworth (1980) has shown that almost all of the Na channels of a node of Ranvier are open at the peak Na current that flows upon a strong depolarization. Thus the normalization procedure is likely to yield fairly correct transient open probabilities, so if different models are discussed, they should be constrained to give approximately these transient open probabilities.

In comparison with the traces in Fig. 3, a model that gives a larger r_{slow} than the Hodgkin-Huxley model then needs to have enhanced openings in the late (stationary) bursts, while maintaining similar early (transient) bursts. A trivial modification to achieve this would be an additional process that has a correlation time longer than that of inactivation and that tends to enhance open probability. If one excludes this possibility, the most likely source of difference between an early and a late group of bursts is the unique initial condition that

precedes the very first burst and makes this burst differ from all
subsequent bursts.

   In the experiment simulated in Fig. 3, the channel was
assumed to start from a non-inactivated state. Channels that
contribute to the transient current are very likely to start in
this way, particularly if inactivation has been removed by a
hyperpolarizing prepulse. In all bursts of openings after the one
that starts right upon the depolarization, the channel is
initially returning from an inactivated state. Open probability
within these bursts could be enhanced with respect to that in the
first one if the return from inactivation is associated with a
temporary increase in open probability. Such an increase is
certain if inactivated and non-inactivated states interconvert
always through an open state (the channel then also opens each
time before it inactivates, but this phenomenon should affect all
bursts to similar extents.). The increase is more subtle if
inactivation occurs through partially activated states as well.
Several observations seem to rule out the strict coupling between
opening and inactivation. Thus, recovery from inactivation is not
associated with a detectable increase in Na current of squid axons
(Bezanilla & Armstrong, 1977). Individual Na channels of rat
myotubes may become inactivated without having opened to
contribute to a transient Na current (Horn et al., 1981; Patlak &
Horn, 1982). On the other hand, the partial immobilization of
gating charge indicates that activation cannot be fully reversed
once the channel is inactivated (Armstrong & Bezanilla, 1977;
Nonner, 1980). Thus the channel is likely to inactivate, or
recover, through at least partially activated states. Conti et
al. (1980) have examined several models of the latter kind when
trying to reproduce the spectra in Fig. 2. These models yield
slow fluctuations that are marginally stronger than those of a
Hodgkin-Huxley model, but their spectral densities still deviate
significantly from the experimental ones.

We are left with the conclusion that the ratio r_{slow} is not likely to vary much among different kinetic models of gating, even among models that differ in the degree of kinetic coupling between the major gating processes. We therefore should consider other possibilities in interpreting the difference between experimental and theoretical power spectral densities.

One such possibility has already been mentioned: the spectra from the small depolarizations in Fig. 2 suggest that the variance contained in the frequencies beyond 5 kHz might be larger than is anticipated by the models considered so far. These models then may predict r_{slow} correctly, but scaling r_{slow} by a total variance that is too small will produce an underestimate of the absolute low-frequency variance as well. This has been, indeed, the major difficulty in reconstructing the experimental spectra in Fig. 2. We may thus examine what kind of activation kinetics could be compatible both with the observed relaxations and a suspected high intensity of fast fluctuations.

FAST FLUCTUATIONS FROM THE ACTIVATION PROCESS

Very fast current fluctuations may well be produced by a gating mechanism that opens relatively slowly in a relaxation experiment. This requires only that the open channel can (partially) close and re-open by quick transitions that are masked by slower transitions when the channel moves all the way from resting to open states. The Hodgkin-Huxley model already incorporates such tendency, so it may again serve as a basis of discussion.

Activation gating by three independent and identical subunits gives rise to three relaxation time constants that are in fixed ratios $\tau_m : \tau_m/2 : \tau_m/3$ (including an independent inactivation subunit

actually splits them further into three pairs of closely similar
time constants). This is easily verified by evaluating

$$p(\text{activ.}) = m^3 = (m_\infty + (m_0 - m_\infty) \exp[-t/\tau_m])^3, \tag{9}$$

which is the usual notation for activation time courses in the
Hodgkin-Huxley model, into

$$p(\text{activ.}) = m^3_\infty + 3m^2_\infty (m_0 - m_\infty) \exp[-t/\tau_m] +$$
$$3m_\infty (m_0 - m_\infty)^2 \exp[-t/(\tau_m/2)] +$$
$$(m_0 - m_\infty)^3 \exp[-t/(\tau_m/3)] \tag{10}$$

Upon a small step depolarization that activates a small fraction
of the available Na channels ($m_0 - m_\infty \ll 1$), the rise of Na current
is dominated by the exponential in τ_m. During a steady
depolarization to the same level, the Na current fluctuations,
however, are dominated by the term in $\tau_m/3$. This is readily
verified by remembering that, in the time domain, the
autocovariance of fluctuations yields the same kinetics as would
be seen in a relaxation experiment that starts with all channels
initially open ($m_0 = 1$); for this initial condition and a small
depolarization ($m_\infty \ll 1$), eqn. (10) predicts that the exponential in
$\tau_m/3$ prevails. As a numerical example, let us consider the 16 mV
experiment of Fig. 2. The transient Na current in this fiber was
measured with the step applied from a conditioning level of 8 mV
($m_0 = 0.09$; $m_\infty = 0.16$); the (absolute) amplitude ratios for the
exponentials in τ_m, $\tau_m/2$, and $\tau_m/3$ were 1:0.44:0.06. The
variances of the corresponding Lorentzian terms in the power
spectrum were predicted to be in the opposite order: 0.11:0.57:1.
If this, or a similar, model is inaccurate in the fast relaxation
term, the inadequacy is likely to be overlooked in the fits to the
rise of Na current, whereas it is likely to manifest itself in the
fits to fluctuation spectra. Evidently, the Hodgkin-Huxley model

already is capable of producing fast fluctuations that do not have an obvious counterpart in macroscopic relaxations.

Improving the fast fluctuation properties of the Hodgkin-Huxley model will at least require modification of the reaction step that is directly associated with the opening and closing of the gate. Fast fluctuations are preferentially seen under conditions when the probability of activation is low, and then are rate-limited by the rate of closure. This can be recognized, in the case of Hodgkin-Huxley kinetics, from the simulation for the 16 mV depolarization shown in Fig. 3: the dominant correlation time of these current pulses is their lifetime, which at a small depolarization is the reciprocal rate of de-activation, $(3\beta_m)^{-1}$. If a model yields fluctuations beyond the corresponding frequency, it must be able to achieve higher rates of closure than the Hodgkin-Huxley model.

We do not yet know what the cut-off frequency of the fast fluctuations actually is, so the discussion of improved models must be left at this qualitative level. The postulated high rate of closure at small depolarizations is in contrast to the low rate of closure at slightly larger depolarizations, which has been measured directly in patch-clamp experiments on myotubes (Patlak & Horn, 1982). This would indicate that channel closure is highly voltage-sensitive as has already been postulated on the basis of gating current experiments (Armstrong & Gilly, 1979).

REFERENCES

Adelman, Jr., W. J., and Palti, Y., 1969, The effects of external potassium and long duration voltage conditioning on the amplitude of sodium currents in the giant axon of the squid, Loligo pealei, J. Gen. Physiol., 54:589.

Armstrong, C. M., and Bezanilla, F., 1977, Inactivation of the sodium channel. II. Gating current experiments, J. Gen. Physiol., 70:567.

Armstrong, C. M., and Gilly, W. F., 1979, Fast and slow steps in the activation of Na channels, J. Gen. Physiol., 74:691.

Bezanilla, F., and Armstrong, C. M., 1977, Inactivation of the sodium channel. I. Sodium current experiments, J. Gen. Physiol., 70:549.

Conti, F., Hille, B., Neumcke, B., Nonner, W., and Stämpfli, R., 1976, Measurement of the conductance of the sodium channel from current fluctuations at the node of Ranvier, J. Physiol., 262:699.

Conti, F., Neumcke, B., Nonner, W., and Stämpfli, R., 1980, Conductance fluctuations from the inactivation process of sodium channels in myelinated nerve fibres, J. Physiol., 308:217.

Dodge, F. A., and Frankenhaeuser, B., 1959, Sodium current in the myelinated nerve fibre of Xenopus laevis investigated with the voltage clamp technique, J. Physiol., 148:188.

Hodgkin, A. L., and Huxley, A. F., 1952, A quantitative description of membrane current and its application to conduction and excitation in nerve, J. Physiol., 117:500.

Horn, R., Patlak, J., and Stevens, C. F., 1981, Sodium channels need not open before they inactivate, Nature, 291:426.

Neumcke, B., Schwarz, W., and Stämpfli, R., 1979, Slow actions of hyperpolarization and sodium channels in the membrane of myelinated nerve, Biochim. Biophys. Acta, 558:113.

Nonner, W., 1980, Relations between the inactivation of sodium channels and the immobilization of gating change in frog myelinated nerve, J. Physiol., 299:573.

Patlak, J., and Horn, R., 1982, Effect of N-bromoacetamide on single sodium channel currents in excised membrane patches, J. Gen. Physiol., 79:33.

Peganov, E. M., Khodorov, B. I., and Shishkova, L. D., 1973, Slow
    sodium inactivation related to external potassium in the
    membrane of Ranvier's node. The role of external K, Bull.
    exp. Biol. Med. U.S.S.R., 25:15.

Sigworth, F. J., 1980, The variance of sodium current fluctuations
    at the node of Ranvier, J. Physiol., 307:97.

Sigworth, F. J., 1981, Interpreting power spectra from
    nonstationary membrane current fluctuations, Biophys. J.,
    35:289.

SYNAPTIC NOISE

Vincent E. Dionne

Department of Medicine/Division of Pharmacology
University of California, San Diego
La Jolla, CA 92093, USA

Synaptic noise as a recognized and useful biological signal
has been with us since Sir Bernard Katz and Ricardo Miledi
described the increased voltage noise in the
acetylcholine-induced endplate potential recorded from a frog
neuromuscular junction (Katz and Miledi, 1970). Noise analysis
has produced important insights into the function of the synaptic
mechanism, and the success of this approach has encouraged
similar undertakings in a broad variety of other biological
systems. In this paper I will discuss several topics about the
interpretation of noise data and its role in the study of
molecular mechanisms. While specific points will be illustrated
using synaptic noise studies, the reader should note the general
applicability of each topic to studies of other biological noise
sources.

Both in this volume and elsewhere the point has been made
that biological noise is a signal that contains information about
the mechanism which produces it. Noise analysis is the endeavor
to understand that information. At the neuromuscular junction
synaptic noise is the net result of a large number of

139

acetylcholine receptors being activated together but
asynchronously (Adams, 1981). Such activation occurs for example
when a cholinergic agonist such as acetylcholine is applied in a
prolonged fashion. That is, if the agonist is exposed to the
receptors for periods which are long compared to the normal time
course of receptor activity, one will see fluctuations in the net
synaptic response because individual receptors will be activated
at random times while their duration of activation will vary.
Over the time course of the agonist application the signal
fluctuates when the number of active receptors changes from
instant to instant. In typical studies at the neuromuscular
junction approximately 10^4 receptors may be simultaneously
active. Although this is a small fraction of the total number of
receptors available (about 10^7 per endplate), it is still a large
number. Clearly then, the properties of receptor function which
can be evaluated from fluctuations must represent the average
behavior of the active population. In this sense, noise analysis
supplements the more direct single channel studies made on the
same receptor system (Patlak, 1984).

Acetylcholine receptors are integral membrane proteins which
exist in high density at the neuromuscular junction. They
mediate synaptic transmission by increasing the endplate membrane
cation conductance in response to nerve-released acetylcholine,
thus producing an endplate potential. The receptor behaves as
though it is comprised of two separate functional entities: a
binding subunit capable of binding two molecules of acetylcholine
or other chemical agonist, and a channel entity which provides a
high conductance path through the membrane for small cations when
the channel is open. Although fluctuations in the membrane
voltage can be induced and recorded in response to exogenously
applied agonists, most contemporary work relies on the
voltage-clamp method to reveal fluctuations in the membrane
conductance. These conductance fluctuations are the summed

response of the activated or open single channels, each a part of
an acetylcholine receptor. It is this signal which can be
analyzed to obtain estimates of the average single channel
conductance parameters.

In a formal sense, all the information contained in a noisy
synaptic response is described by three parameters, each of which
may vary with time:
  a) the mean value of the measured signal
  b) the total variance of the signal around its mean value
  c) the frequency distribution of the variance (the spectral
     density function).
Thus, to study the cholinergic conductance mechanism using noise
analysis, these three parameters must be estimated. The mean
value of the induced conductance change is taken as the
difference between the resting membrane conductance when no
agonist bathes the receptors and the conductance in the presence
of a maintained concentration of agonist. It is sufficient to
average over several tens of milliseconds to obtain the mean
value because the characteristic time constant of the noise
varies between 1 and 10 msec typically. The total variance
relative to the mean may be obtained in several ways. For
example, one could square the output of an RMS meter which had
been set to monitor the AC-coupled signal. Alternatively and
more practically, the total variance may be computed by
integrating the spectral density function over its frequency
range (Bendat and Piersol, 1980). The spectral density function
is simply an accounting of how the fluctuation intensity or
variance depends upon frequency. For synaptic noise the spectral
density at low frequencies is more or less constant, and it
decreases progressively at higher frequencies. The spectral
density function for any particular data record may be obtained
with a direct Fourier transformation of the record itself.

Useful descriptions of the technique are to be found in Neher and Stevens (1977), Lecar and Sachs (1980), and Dionne (1981).

The three parameters which formally describe the fluctuations of the conductance signal may all depend on time. When they do change with time, one must deal with the problem of nonstationary noise analysis. In principle, nonstationary noise requires much greater computational sophistication for its analysis (1981); when possible, experimental conditions are arranged so that stationary (time invariant) noise parameters may be estimated. As a rule, synaptic noise studies deal with quasi-stationary data, and it is assumed for the purposes of analysis that the data is strictly stationary.

The interpretation of noise data is a completely separate endeavor from the activity of estimating the descriptive parameters for a particular record. Interpretation requires a comparison between predictions derived from hypothetical models of the molecular mechanism and the noise characteristic parameters: the mean value, total variance, and spectral density function. Since the synaptic noise data record is an accounting for how the number of active receptors changes with time, useful models must describe the kinetic behavior of receptors. For this purpose mass action models have been widely applied because they are readily treated mathematically to predict the noise characteristics. However, with a modeling approach for the interpretation of noise data there can be no unique description of the molecular mechanism. This follows because one can always imagine kinetic schemes of increasing complexity which presumably simplify under the measurement conditions to fit the data. This is not an indictment of noise analysis but simply points out the limitations inherent in any model-building endeavor. The best that can be expected in the way of uniqueness is to identify the simplest kinetic model necessary to describe the data

characteristics. Although an interpretation of mechanism depends
upon the model employed, on occasion, model-independent features
can be identified.

How then can the characteristic noise parameters be
predicted from any particular kinetic model. These parameters
are the result of a statistical mathematical treatment, the
details of which can be found in a number of texts. An
excellent, comprehensive reference is Bendat and Piersol (1980);
briefer outlines focussed on synaptic noise are contained in the
technical references given above. Below I will summarize the
important concepts. Recall that we are considering here the
measurement of membrane conductance, and that each receptor is
thought to provide a modest increment in that measured
conductance when the receptor is active. In thinking about a
kinetic model to describe the transient behavior of the receptor,
one or more states must be identified as active or conducting
while the other(s) will be inactive or nonconducting. The
statistical treatment is then concerned with several issues:
a) What fraction of time is the receptor in the active state?
b) How frequently does the receptor become active? c) How long
does a receptor remain active? In sum, we want to obtain a
prediction of the time dependence of the probability that a
typical receptor is active - given the experimental conditions.
This is achieved using the probability density function, which is
derived from and depends uniquely upon the particular kinetic
model being treated. With the probability density function one
can compute the mean value for the probability that the receptor
resides in the active state and the covariance function of that
probability. The covariance function characterizes the
fluctuations observed in the predicted probability that the
typical receptor is active. It contains the same kind of
information as the spectral density function except that the
covariance is a function of time while the spectral density

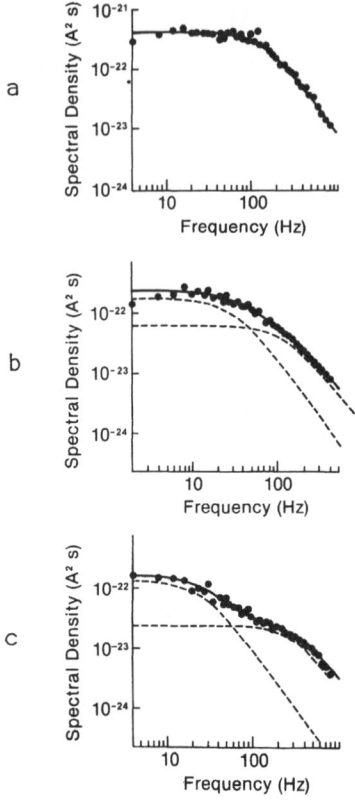

Fig. 1.   Three different examples of synaptic noise spectra. The
         data used to produce these spectra were recorded from
         garter snake neuromuscular junctions using a voltage
         clamp and standard noise techniques. Note that these
         are current fluctuation spectra, while in the text
         synaptic noise is discussed in terms of conductance
         fluctuations. Because the voltage clamp holds the
         membrane potential (V) constant, current (I) is
         proportional to conductance (g) according to $I(t) =
         g(t)[V - V_r]$ where V_r (the reversal potential) is
         approximately -3 mV.

         (a) Carbamylcholine induced noise spectrum from twitch
         muscle endplate, -79 mV membrane potential. $S(0) =$

function depends upon frequency. The spectral density function
may be obtained by Fourier transforming the covariance function.
The total variance is obtained either as the intercept of the
covariance function at t = 0 or as the integral over frequency of
the spectral density function.

Synaptic noise data contain several features which must be
accounted for by the kinetic models. Of major importance are the
characteristic frequencies or rates which describe the net
transitions between active and inactives states of the receptor.
The presence of characteristic frequencies in the synaptic data
is illustrated in Figure 1. They appear in the log-log spectral
density plots as limiting frequencies beyond which the amplitude

4.0×10^{-22} A^2sec, f_o = 150 Hz, γ = 39.9 pS (assuming
$\beta' \ll \alpha$).(b) Acetylcholine induced noise spectrum from
a slow (tonic) muscle endplate at −61 mV fitted with
two components as described in the text. $S_1(0)$ =
2.44×10^{-22} A^2sec, f_1 = 35 Hz; $S_2(0)$ = 8.75×10^{-23}
A^2sec, f_2 = 144 Hz. These values give (according to
Model B in the low concentration limit) α = 600 sec^{-1},
β = 170 sec^{-1}, k_2 = 300 s^{-1} and an adjusted single
channel conductance at 25.2 pS.
(c) A synthetic agonist-induced noise spectrum from a
twitch muscle endplate at −114 mV fitted with two
components. $S_1(0)$ = 1.58×10^{-22} A^2sec, f_1 = 30 Hz;
$S_2(0)$ = 2.85×10^{-23} A^2sec, f_2 = 396 Hz. Because the
interpretation of noise data is model-dependent while
the mechanism which produces two-component synaptic
noise here is not known, estimates from this data alone
of single channel properties are not possible.

falls rapidly. Depending upon the actual features of the
mechanism, there may be one or more characteristic frequencies
and not all of them may be clearly resolved in the data. Kinetic
models predict that such characteristic frequencies as these
should be observed; the prediction is that each characteristic
frequency appear as the "corner frequency" of a Lorentzian
component in the spectral density function. A Lorentzian
function relating spectral density to frequency has the following
form:

$$S(f) \; = \; \frac{S(0)}{1 + \left(\dfrac{f}{f_o}\right)^2}$$

where $S(0)$ is the limiting spectral density when $f = 0$ and f_o is
the "corner" or characteristic frequency. The theoretical line
through the data in Figure 1a is a single Lorentzian function;
the Lorentzian components of the other spectra in Figure 1 are
drawn as dotted lines the sum of which is displayed as the smooth
line through the data points. It is most important not to
confuse the characteristic frequencies and the transition rates
specified in a kinetic model. That is, the corner frequencies of
the Lorentzian components in Figure 1 cannot be identified simply
as specific transition rate constants in whatever kinetic model
was used to generate the theoretical curves. As will be
illustrated below, the characteristic frequencies are functions
of all the kinetic transition rate constants. Another point will
be documented in the following examples although not proved in
general. That is, a kinetic model predicts a number of
characteristic frequencies which is one less than the number of
kinetic states in the model. Thus, a two-state model will give
one characteristic frequency (or equivalently one Lorentzian
component in the spectral density function) and a three-state
model will give two.

Let us now consider the data in Figure 1, see how the theoretical curves drawn there were obtained and fitted, and evaluate the interpretations which may be drawn. The data in Figure 1a was recorded from the neuromuscular junction of a garter snake (sp. Thamnophis) twitch muscle fiber using the synthetic agonist carbamylcholine. Over the frequency range of 4-1000 Hz it was well fitted by a single Lorentzian component. Similarly shaped spectra have been obtained using acetylcholine at these garter snake neuromuscular junctions (Dionne and Parsons, 1981) and at other vertebrate neuromuscular junctions [e.g. frog (Anderson and Stevens, 1973) and rat (Sachs and Lecar, 1977)].

To generate a theory that could fit this data it was necessary to use only a two-state kinetic model with one state representing the active, open channel and the other the closed channel, both with bound agonist molecules.

$$\text{closed} \underset{\alpha}{\overset{\beta'}{\rightleftarrows}} \text{open} \qquad \text{(A)}$$

Here the characteristic frequency will be seen to describe kinetic changes between these open and closed states. The particular frequency is characteristic in the sense that it generally applies for all perturbations employed to stimulate a change in distribution among the states; it is a characteristic of the model. One way to calculate it is to imagine that at some initial time the closed channel population is suddenly raised from zero to N and compute the time course by which equilibrium is established between the states. Let us do that here by setting $\eta_o(t)$ = the number of open channels and $\eta_c(t)$ = the number of closed channels so that $\eta_o + \eta_c = N$; initially $\eta_o(0) = 0$, $\eta_c(0) = N$. The rate of change of $\eta_o(t)$ may be written according to model A

$$\frac{d\eta_o(t)}{dt} = \beta'\eta_c(t) - \alpha\eta_o(t)$$

or, substituting for $\eta_c(t)$,

$$\frac{d\eta_o(t)}{dt} = \beta'N - (\alpha + \beta')\eta_o(t)$$

The solution of this differential equation is easily obtained by Laplace transform methods; by substitution it can be confirmed that

$$\eta_o(t) = \frac{\beta'N}{\alpha + \beta'} [1 - e^{-(\alpha + \beta')t}] \qquad (1)$$

Notice that the agonist concentration dependence of $\eta_o(t)$ is not made explicitly clear by the model although it is the exposure to agonist which causes these channels to open. Below we will consider a more complete model of receptor activation which takes this into account. However, let us first examine the response predicted by equation (1) more closely.

The characteristics rate of model A is $\omega_o = (\alpha + \beta')$, which appears as the exponential rate in equation (1). This may be expressed as a characteristic frequency.

$$f_o = \frac{(\alpha + \beta')}{2\pi}$$

This characteristic frequency will be the "corner frequency" of a single Lorentzian component which Model A predicts as the spectral density function. Two slightly different approaches for obtaining the spectral density function of conductance fluctuations from this two-state model can be found in Lecar and Sachs (1980) and Dionne (1981). The result is

$$S(f) = \frac{4N\gamma^2\alpha\beta'/(2\pi f_o)^3}{1 + \left(\dfrac{f}{f_o}\right)^2} \qquad (2)$$

where γ is the conductance of one open channel (active state). This is a single-sided expression for the spectral density function, appropriate for $f > 0$. It has been drawn as the theoretical curve in Figure 1a. Recall that equation (2) describes the redistribution of variance with frequency; the total variance of the conductance fluctuations (σ^2) may be obtained as its integral:

$$\sigma^2 = \int_0^\infty S(f)df = \frac{N\gamma^2\alpha\beta'}{\alpha + \beta'}$$

In addition, the mean conductance (μ) for a stationary response ($t \to \infty$) may be written according to equation (1) as

$$\mu = \gamma\eta_o(\infty) = \frac{N\gamma\beta'}{\alpha + \beta'}$$

The ratio of variance to mean provides an expression for the estimation of γ, the mean single channel conductance.

$$\frac{\sigma^2}{\mu} = \gamma\left(\frac{\alpha}{\alpha + \beta'}\right)$$

Then, if $\alpha >> \beta'$ under the experimental conditions, this ratio will
give a rather accurate measurement of single channel conductance.
Arguments have been advanced that α is much greater than β' for
receptors at several vertebrate neuromuscular junctions, but
frequently this difference is assumed (Andersen & Stevens, 1973;
Neher & Stevens, 1977). Under many experimental conditions where
fluctuations are produced with agonist concentrations that are
low with respect to their equilibrium dissociation constants and
the spectral density is well described by the two state model,
these assumptions seem to be valid. For example, the spectrally
derived estimate of mean single channel conductance using the
data from Figure 1a is 39.9 pS. Direct measurements using the
patch recording technique on carbamylcholine-stimulated endplate
acetylcholine receptors in the same cell type and under very
similar experimental conditions gives 42.0 pS (M.D. Leibowitz,
personal communication).

A more plausible model of receptor kinetics can be written
which specifically includes agonist binding; the model uses three
kinetic states, although the spectral density function in Figure
1a does not necessitate a three-state model.

$$2A + R \underset{k_2}{\overset{k_1}{\rightleftarrows}} (A_2R) \underset{\alpha}{\overset{\beta}{\rightleftarrows}} A_2R* \qquad \text{(B)}$$

Consider the sequential scheme above where two molecules of
agonist A can bind simultaneously to a closed channel – receptor
R producing the closed channel state (A_2R) and the open channel
state A_2R*. This three-state model will provide two
characteristic frequencies or rates which reflect the kinetics of
channel distribution among the three states. The characteristic
rates may be computed in a similar fashion as earlier. Let $\eta_o(t)$
= the number of open channels A_2R*, $\eta_c(t)$ = the number of closed

channels (A_2R), and N = the total number of channels. Suppose that the response is initiated when the free agonist concentration is suddenly changed from zero to a constant, non-zero value = C. Two differential equations describing separately the rates of change of $\eta_o(t)$ and $\eta_c(t)$ can be written according to Model B.

$$\frac{d\eta_o(t)}{dt} = \beta\eta_c(t) - \alpha\eta_o(t)$$

$$\frac{d\eta_c(t)}{dt} = k_1C^2[N - \eta_o(t) - \eta_c(t)] + \alpha\eta_o(t) - (\beta + k_2)\eta_c(t)$$

These equations must be solved simultaneously (e.g., by Laplace methods) to estimate the time course of the open channel population; the solution is

$$\frac{\eta_o(t)}{N} = \frac{k_1C^2}{\omega_1\omega_2} - \frac{k_1C^2 e^{-\omega_2 t}}{\omega_2(\omega_1-\omega_2)} + \frac{k_1C^2 e^{-\omega_1 t}}{\omega_1(\omega_1-\omega_2)}$$

According to this description the number of open channels changes as the sum of two exponential components, approaching a limiting maximum of $Nk_1C^2/\omega_1\omega_2$ at long times.

The two characteristic rates are

$$\omega_1 = [(\tfrac{1}{2}(\alpha + \beta + k_2 + k_1C_1^2)$$
$$_2$$
$$\pm [(\alpha + \beta + k_2 + k_1C_1^2)^2 - 4\alpha k_2 - 4(\alpha + \beta)k_1C^2]^{\frac{1}{2}}]$$

where the subscript 1 or 2 on ω_1 corresponds to the + and - sign
$$_2$$

respectively on the right-hand side of the equation. As with the
two-state model, the characteristic rates depend explicitly upon
all of the transition rate constants in the model. Moreover,
these rates are seen to depend upon agonist concentration; as C
increases, both ω_1 and ω_2 will increase. Model B with its two
characteristic rates predicts a spectral density function which
is the sum of two Lorentzian components.

The predictions of Model B will reduce to those of the
earlier two-state scheme and thus appear consistent with the data
in Figure 1a if it is assumed that the transition rate constants
accounting for agonist binding and unbinding ($C^2 k_1$ and k_2) are
very large relative to the rate constants for channel closing and
opening (α and β). By this approximation the characteristic
rates reduce to

$$\omega_1 = k_2 + k_1 C^2$$

$$\omega_2 = \alpha + \beta$$

where ω_1 has been assumed to be much bigger than ω_2. The corner
frequency seen in Figure 1a should then represent $\omega_2/2\pi$ because
$f_1 = \omega_1/2\pi$ must be such a high frequency that no change in the
spectral density over the frequency band shown would be observed
from its Lorentzian component. Thus, by assuming that a
relatively large disparity exists between the magnitudes of the
binding and conformational rates, the predictions of Model B
simplify to resemble those of Model A. This step is taken
because only two states can be effectively discriminated in
Figure 1a with the measurement resolution available.

An expression for the spectral density function predicted by
Model B may be obtained in a manner similar to that illustrated
for the two-state model. A more general approach is given by

Colquhoun and Hawkes (1977). The result for the predicted conductance spectrum is (Dionne, 1981a)

$$S(f) = \frac{S_1(0)}{1 + \left(\frac{2\pi f}{\omega_1}\right)^2} + \frac{S_2(0)}{1 + \left(\frac{2\pi f}{\omega_2}\right)^2}$$

where

$$S_1(0) = \frac{4NC^2\beta\gamma^2 k_1}{k_2\alpha d^2(\omega_1 - \omega_2)} \cdot \frac{\omega_2(\omega_2 - \alpha)}{\omega_1\omega_2 - k_1 C^2}$$

$$S_2(0) = \frac{4NC^2\beta\gamma^2 k_1}{k_2\alpha d^2(\omega_1 - \omega_2)} \cdot \frac{\omega_1(\alpha - \omega_1)}{\omega_1\omega_2 - k_1 C^2}$$

and

$$d^2 = \frac{C^2 k_1(\alpha + \beta)}{k_2\alpha} + 1$$

Although the two spectral components are additive, neither is identical to the single component obtained from the simpler two-state model.

In synaptic noise studies care is usually taken to insure that agonist concentration is low relative to the equilibrium dissociation constant between agonist and receptor. There are several reasons. (a) In this limit receptor desensitization (loss of activity produced by prolonged exposure to agonist) is a minor problem, so that nearly stationary noise data may be recorded and analysed. (b) Only a small fraction of receptors will be active at any time so that the total noise variance remains proportional to the mean induced response and an unbiased estimate of single channel conductance can be obtained. (c) Only small induced currents are produced which are experimentally easier to control and better tolerated by the tissue. In this low concentration limit, $c^2 << k_2/k_1$, the characteristic rates reduce to

$$\omega_1 = [\tfrac{1}{2} (\alpha + \beta + k_2) \pm [(\alpha + \beta + k_2)^2 - 4\alpha k_2]^{\tfrac{1}{2}}] \quad (5)$$

That is, the concentration dependence of the characteristic rates is lost at low agonist concentrations; however, this three-state model continues to predict two characteristic rates. If the transition rate constants α, β, and k_2 had values such that $\omega_1/2\pi$ and $\omega_2/2\pi$ fell within the detected frequency bandwidth of the data, one would expect to find spectral density measurements which required two Lorentzian components to be accurately fitted.

This version of the three-state model has been applied in the case of acetylcholine induced synaptic noise data recorded from garter snake slow muscles (Dionne, 1981a; Dionne & Parsons, 1981). An example of a spectrum is shown in Figure 1b. These noise data cannot be fitted with a single Lorentzian component because the variance at higher frequencies is larger than predicted. In this figure the two components have been selected using more information than just that provided by the noise. Specifically, the observed rate of decay of the average miniature endplate current recorded from this cell under conditions identical to those for the noise data was used to constrain the lower frequency component, because, according to Model B, these should be identical. Without such a constraint there would be an intolerably large uncertainty in the relative difference between the two characteristic frequencies which were fitted. For these slow muscle synaptic noise spectra, the transition rate constants α, β, and k_2 which appear in the three-state kinetic model can be estimated using the two characteristic frequencies and the relative amplitude of the two Lorentzian components. Note again that these kinetic transition rate constants cannot be equated

with either of the observed characteristic rates. It would be
incorrect to argue here that the decay of the miniature endplate
current or a particular corner frequency reflected solely the
channel closing rate α, for instance. Instead, the
experimentally observed characteristic rates depend upon all the
transition rate constants in the underlying kinetic scheme. The
mean conductance of acetylcholine receptor channels in slow
muscle may be obtained from the ratio of total variance to mean
induced conductance, just as described above. In the low
concentration limit the relation between total variance, mean
conductance and γ reduces to equation 3 with β replacing β'.
However, this ratio will give only a fraction of the true single
channel conductance because here β is not negligible compared to
α. Since both α and β can be evaluated from the noise analysis,
the estimate of single channel conductance from the ratio σ^2/μ
can be corrected (Dionne, 1981a).

It is interesting finally to examine Figure 1c; it shows a
noise spectrum from data recorded when a synthetic agonist, an
NBD acylcholine, [4-(7-nitrobenzo-2-oxa-1,3-diazole) acylcholine
prepared by M. Bolger] was applied to the neuromuscular junction
of a garter snake twitch muscle. The experimental situation was
quite similar to that which produced the data in Figure 1a: an
artificial agonist on a twitch muscle endplate. Over the
recording bandwidth of 4-1000 Hz the spectrum clearly showed two
Lorentzian components, so that the minimal model of synaptic
activity must have three kinetic states. However, it would be
premature to fit this spectrum with the predictions from the
three-state model used for slow muscle synaptic noise because
that model makes a very specific assignment of the molecular
process which produces the three kinetic states. Without
additional evidence it is not clear what the process is that
underlies this noise spectrum. One quite plausible alternative
would be that the synthetic chemical agonist can also behave like

a local anesthetic, producing an effective blockade of the open
channel, similar to decamethonium (Adams & Sakmann, 1978). A
scheme for such a mechanism with three kinetically
distinguishable states would be an extension of the two-state
model (A) originally introduced:

$$\text{closed} \underset{\alpha}{\overset{\beta}{\rightleftharpoons}} \text{open} \underset{b_-}{\overset{b_+}{\rightleftharpoons}} \text{blocked} \qquad (C)$$

Without having access to more information that just that provided
by the noise spectrum, selection of a mechanism would not be
possible.

In summary, although all three spectral density functions in
Figure 1 are from synaptic noise studies, each shows a different
dependence of variance on frequency. Notwithstanding this, each
can be fitted with the same three-state kinetic model (B). The
rationale for fitting the simple data in Figure 1a with the
three-state model was not one of necessity. Rather, it was
argued that more than a two-state model was needed to account for
the agonist concentration dependence of the evoked conductance.
The result was that simplifying assumptions were required to
reduce the predicted spectral density function to a single
resolved component. The method chosen is not unique but is
believed to represent accurately the essential underlying
molecular behavior of receptors in twitch muscle synapses. In
contrast, a three-state model is necessary to fit the spectra in
Figures 1b and c, because at least two Lorentzian components are
needed in each. Model B is believed to represent an accurate
minimal kinetic relation for the behavior of garter snake slow
muscle cholinergic receptors, Figure 1b (Dionne, 1981a). Of

course Model B must be a simplification of a more accurate kinetic description of the receptor since the precisely simultaneous binding and unbinding of two agonist molecules is rather implausible and other functional behaviors of the receptor such as desensitization have been ignored. Nevertheless, it should be noted that noise data alone was not enough to identify Model B as the better choice from among mechanisms represented by other three-state models. The kinetics of synaptic responses other than noise were also analysed, and the requirement of internal consistency among several kinetic responses was used to validate model B.

In general, noise will not provide enough information on which to base the identification of a most likely molecular mechanism. Noise can be a most useful piece of data to analyse, but it must stand together with all the other data that can be obtained from a system. Figure 1c illustrates this same point, for, although it requires a three-state kinetic model to be fitted, without knowledge of how other types of kinetic responses behave one has no reason to select the molecular mechanism represented by Model B over that of many other mechanisms, for example Model C. Noise analysis can help to unravel the molecular properties of many biological responses, but it cannot serve alone.

ACKNOWLEDGEMENTS

The author acknowledges support from the National Institutes of Health (NS 15344) and the Office of Naval Research (N00014-79-C-079).

REFERENCES

Adams, P. R., and Sakmann, B., 1978, Decamethonium both opens and
    blocks endplate channels, Proc. Natl. Acad. Sci. USA,
    75:2994.

Adams, P. R., 1981, Acetylcholine receptor kinetics, J. Membrane
    Biol., 58:161.

Anderson, C. R., and Stevens, C. F., 1973, Voltage clamp analysis
    of acetylcholine produced end-plate current fluctuations at
    frog neuromuscular junction, J. Physiol., 235:655.

Bendat, J. S., and Piersol, A. G., 1980, "Engineering
    Applications of Correlation and Spectral Analysis," John
    Wiley and Sons, Inc.

Colquhoun, D., and Hawkes, A. G., 1977, Relaxation and
    fluctuations of membrane currents that flow through drug-
    operated channels, Proc. R. Soc. Lond. B., 199:231.

Dionne, V. E., and Parsons, R. L., 1981, Characteristics of the
    acetylcholine-operated channel in twitch and slow fiber
    neuromuscular junctions of the garter snake, J. Physiol.,
    310:145.

Dionne, V. E., 1981a, The kinetics of slow muscle acetylcholine-
    operated channels in the garter snake, J. Physiol., 310:159.

Dionne, V. E., 1981b, Noise analysis in "Techniques in Cellular
    Physiology," P. F. Baker, ed., Elsevier/North-Holland
    Biomedical Press.

Katz, B., and Miledi, R., 1970, Membrane noise produced by
    acetylcholine, Nature, 226:962.

Lecar, H., and Sachs, F., 1980, Membrane noise analysis, in
    "Excitable Cells in Tissue Culture," P. G. Nelson and
    M. Liebennau, ed., Plenum Publishing Corp., New York.

Neher, E., and Stevens, C. F., 1977, Conductance fluctuations and
    ionic pores in membranes, Ann. Rev. Biophys. Bioeng., 6:345.

Patlak, J., 1984, The information content of single channel data
    in, "Membrane, Channels and Noise, R. S. Eisenberg, M.
    Frank, and C. F. Stevens, ed., Plenum Publishing Corp., New
    York.

Sachs, F., and Lecar, H., 1977, Acetylcholine-induced current
    fluctuations in tissue-cultured muscle cells under voltage
    clamp, Biophys. J., 17:129.

Sigworth, F. J., 1981, Interpreting power spectra from
    nonstationary membrane current fluctuations, Biochem. J.,
    35:289.

NOISE ANALYSIS OF TRANSPORT THROUGH APICAL SODIUM CHANNELS

OF TIGHT AMPHIBIAN EPITHELIA

Bernd Lindemann and Jack H.-Y. Li

2nd Department of Physiology
University of the Saarland
D-6650 Homburg/Saar, FRG

Among the pecularities of epithelial tissues two are of
special concern for noise analysis: the complex architecture and
the multitude of transport regulating mechanisms.

The architecture, although otherwise quite variable among
epithelia of different origin, is always based on two membranes
in series (the apical and the basolateral membrane), which
delimit the transcellular transport pathway, and on a
complementary paracellular 'shunt' pathway. A given transport
process, therefore, has to be localized in one or more of these
three elements, and the remaining elements usually complicate a
precise biophysical characterization of the process. Here much
advance may be expected from methods of spacial isolation
(vesicle preparations, patch clamp).

The transport regulating mechanisms, which must be studied
in the living cell, or at least also in the living cell, may be
divided into: a) Those which ensure a concerted performance of
the two membranes and survival of the epithelial cell. They
generally protect cellular or 'local' interests; and b) Those

161

which serve 'public' interests by adapting transepithelial
transport to the needs of the organism.

Noise analysis of epithelial Na transport at the apical
membrane has so far dealt with the architectural complexity by
the use of specific blockers like amiloride and by partial
elimination of the basolateral membrane. Regulatory processes of
the local and the public variety have been studied. In both
cases the variation in channel density appears to be the dominant
principle. Since noise analysis affords a nondestructive way to
assay channel density, it is particularly suited for the
investigation of Na transport regulation.

Noise analysis is typically a kinetic method used in
conjunction with other techniques. This article will briefly
survey some relevant features of Na transport in tight epithelia
as worked out with methods other than noise analysis. Against
this background the results of noise analysis are discussed.

ELECTROPHYSIOLOGY OF APICAL NA TRANSPORT IN FROG SKIN

In their fundamental two-membrane-model for Na transport in
frog skin, Koefoed-Johnsen and Ussing (1958) emphasized the
dominant Na permeability of the apical membrane and the dominant
K permeability of the basolateral membrane. The basolateral
membrane was also recognized to contain the Na-K pump (Fig. 1).
As a result of this distribution of transport properties
transcellular Na movement becomes possible, while K ions cycle
through the basolateral membrane. It is remarkable that thereby
the cellular Na-concentration can remain relatively small
although the Na throughput is large.

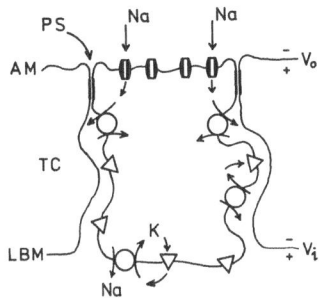

Fig. 1.   Schematic cross-section through a granular epithelial
          cell, showing the essential features of the
          Koefoed-Johnsen Ussing model. AM: apical membrane; LBM:
          basolateral membrane; TC: transport compartment; PS:
          paracellular shunt. V_o, V_i: voltage of apical and
          basolateral membrane. Amiloride blockable Na channels
          are found predominantly in the apical membrane and
          K-channels (triangles) as well as Na,K pumps in the
          basolateral membrane, facing the interstitial
          compartments.

When the overall steady state Na transport is plotted
against the mucosal Na activity (Na_o), a saturating function
resembling a Michaelis-Menten relationship is usually found
(e.g., Ussing, 1949). With two membranes in series many effects
may cause or contribute to this behavior. It was recognized
early, however, that a saturation of active transport is not
involved; rather, the apical permeability, P_{Na}, decreases with
increasing Na_o (Cereijido et al., 1964). This, in turn, may be
caused by a saturation of the apical transport step itself, or be
achieved by a control mechanism. This question was studied with
fast concentration changes of the mucosal solution (Lindemann and
Gebhardt, 1973; Fuchs et al., 1977). When Na_o was suddenly
increased at nearly constant membrane voltage, the Na current

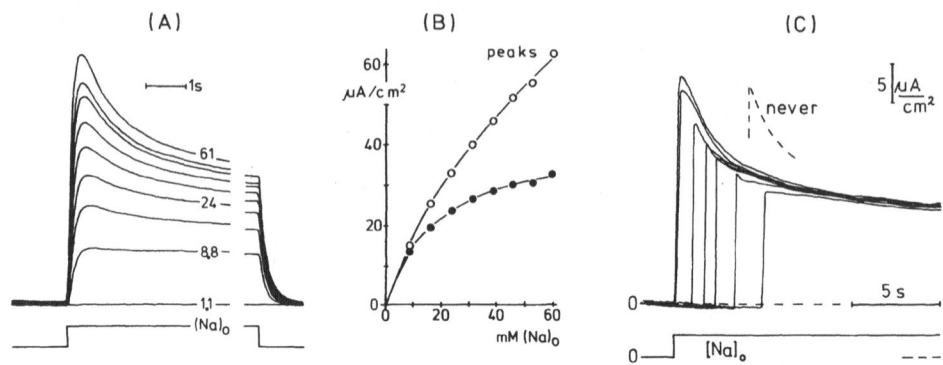

Fig. 2.   Response of I_{Na} (K-depolarized frog skin) to increasing
steps of Na_o in a fast flow chamber, which lowers
unstirred layers to 12 - 15 μm, including the Str.
corneum. (A) Time-course with Na_o indicated as
parameter.   (B) Plot of peak and plateau currents from
panel A against Na_o. The difference between the two
curves estimates the extent of self-inhibition.   (C)
When Na flow is prevented in the early phase of the
response, the self-inhibition develops nevertheless,
showing that it does not depend on electrogenic inflow
of Na ions (Zeiske and Lindemann, see Lindemann, 1977).

rose sharply, passed through a maximum and then settled down to
its much smaller steady state value with time constants in the
order of seconds (Fig. 2A). The steady state currents, when
plotted against Na_o, followed the familiar saturating curve, but
the larger peak currents had a much smaller tendency to saturate
(Fig. 2B). These experiments showed rather clearly that the
apical membrane can pass, although transiently, much more current
than the saturating steady state function indicates. Therefore,
the saturation cannot be caused by an absolute rate limitation of
the transport process itself. The remaining possibility is a

regulatory process, which may be set in effect by the increase in Na_c or the increase in Na_o.

To decide this question the Na reversal potentials F_{Na} were measured before and after the concentration change, and the time course of Na_c computed from them. It turned out that the increase in Na_c was very moderate in the first few seconds during which I_{Na} decreased (Fuchs et al., 1977). Furthermore, the time course of I_{Na} remained essentially unchanged when Na entry was prevented, in the early period of the Na-exposure, by appropriate manipulations of the holding potential (Fig. 2C). By these experiments an increase in Na_c was excluded as being instrumental in the regulatory process.

It was concluded, therefore, that extracellular Na ions act as effectors which decrease P_{Na} with a first order delay of several seconds. The process, which may be called Na self-inhibition, was studied further with near-instantaneous current voltage curves, which showed that for a voltage range of about 60 mV the I_{Na} V-relationship of the apical membrane can be described by the Goldman-Hodgkin-Katz (GHK) equation (Fig. 3). P_{Na} (and Na_c) were estimated by curve fitting and it was found that P_{Na} decreases with time at near constant Na_c. Furthermore, it was confirmed that the steady state P_{Na} decreases with Na_o at near constant Na_c (Fuchs et al., 1977).

In these experiments the membrane voltage was kept near zero by exposing the basolateral membrane to a high K concentration, a device previously used by others (e.g., Rawlins et al., 1970; Morel and Leblanc, 1975). More recent micro-electrode recordings from frog skin appear to show that the depolarization is not complete (Fisher and Helman, 1981). However, similar measurements with other epithelia, which may be more suitable for

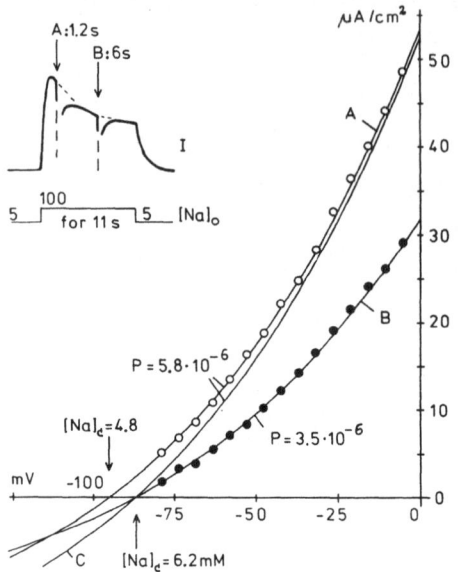

Fig. 3. I_{Na}-voltage curve of apical membrane (K-depolarized frog
skin) recorded during a fast flow experiment at the
times indicated (inset). The continuous curves are fits
with the GHK-equation. Na_c is calculated from E_{Na}
(arrows). P_{Na} is seen to decrease with time. The
indicated values (P) obtained from the fit are in cm/s
(data from Fuchs et al., 1977).

microelectrode work, have shown that the K-depolarization is
reasonably effective and that the IV-curve is indeed of the
GHK-type (Palmer et al., 1981; Frömter et al., 1981; Thompson et
al., 1982; Thomas et al., 1982). Why this result could not be
obtained with microelectrode recordings from frog skin epithelium
is presently unknown. (Our own group discouraged the use of this
technique with frog skin epithelium after noting several
disturbing complications (Nelson et al., 1978).)

It was recognized rather early that in parallel to the
self-inhibition there may be another regulatory process which
also decreases P_{Na} but requires a significant increase in Na_c
(see Lindemann and Gebhardt, 1973, page 126.) Evidence for this
process, which today is called feedback inhibition (Taylor and
Windhager, 1979) first emerged from the work of MacRobbie and
Ussing (1961) and Leblanc and Morel (1975). Grinstein and Erlij
(1978) recognized that basolateral Na/Ca exchange is involved in
this mechanism, as later confirmed by Chase and Al-Awqati (1981,
1982).

To separate these two regulatory processes it is best to
study self-inhibition at constant Na_c, and feedback inhibition at
constant Na_o. In some epithelia, the self-inhibition can be
removed to a large extent by the presence of benzimidazolyl-
guanidine (BIG) in the mucosal solution (Zeiske and Lindemann,
1974). This substance is effective in frog skin but not in toad
bladder. Similarly effective both in skin and bladder are PCMB
and PCMPS (e.g. Dick and Lindemann, 1975; Harms and Fanestil,
1977). However, these compounds may in addition interfere with
the feedback inhibition (Bevevino and Lacas-Vieira, 1982). The
feedback inhibition in toad urinary bladder was recently shown to
have a relatively low inhibitory rate and to be essentially
removed by lowering the serosal Ca concentration to 3 µM (Garty
and Lindemann, 1984).

The availability of amiloride as a potent reversible
inhibitor of apical Na-transport has been available for analysis
both on the membrane level and the molecular level. The
macroscopic inhibition constant of this agent, K_A^{ma} (see equation
12), is in some epithelia increased by Na_o (e.g. Cuthbert and
Shum, 1974; Sudou and Hoshi, 1977; Aceves and Cuthbert, 1979;
Bindslev et al., 1982) suggesting that competition between at
least one type of Na-dependent inhibition and amiloride takes

place. Since the competition phenomenon is found in fast flow
experiments with Na_o exposure times which prevent substantial
increases of Na_c and thereby the feedback inhibition, it is clear
that competition between amiloride and the self-inhibition
exists. However, it may be expected that self-inhibition and
feedback inhibition have interesting cross effects which are to
be studied in the future.

In view of the results to be discussed below, which show
that apical Na-transport is mediated by channels, the competition
phenomenon may be formulated in terms of channel densities. With
the index 0 designating conducting channels, 1 channels blocked
by self-inhibition and 2 channels blocked by amiloride, we may
write the overall reaction scheme

$$N_1 \underset{k_{10}}{\overset{\begin{array}{c} Na_o \\ \downarrow \; k_{01} \end{array}}{\rightleftharpoons}} N_o \underset{k_{20}}{\overset{\begin{array}{c} A_o \\ \downarrow \; k_{02} \end{array}}{\rightleftharpoons}} N_2$$

for pure rather than mixed competitive inhibition. The rate
constants are indicated in this scheme. The dissociation
constants are

$$K_N = k_{10}/k_{01}, \qquad K_A = k_{20}/k_{02}.$$

The sum

$$N = N_o + N_1 + N_2$$

will be called the density of electrically detectable channels.
The steady state probabilities to find a channel in one of the
three states are given by

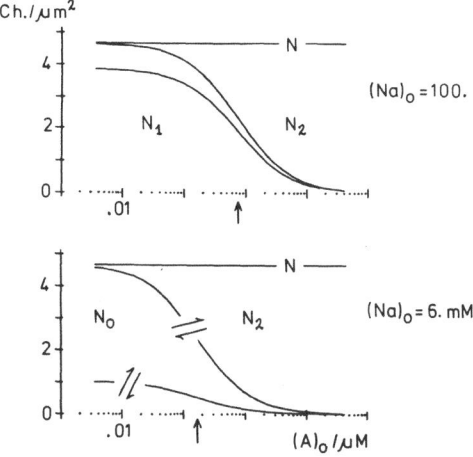

Fig. 4.   Theoretical macroscopic dose response curves of
amiloride, based on eq. 1.   Channel densities are
plotted against log A_o for two values of Na_o.   In each
diagram, the lower S-shaped curves depict N_1, the upper
S-shaped curves N_o plus N_1.   The vertical arrows
indicate K_A^{ma}.

$$P_0 = (1 + Na_o/K_N + A_o/K_A)^{-1}$$

$$P_1 = (Na_o/K_N)\, P_0$$

$$P_2 = (A_o/K_A)\, P_0 \tag{1}$$

The resulting relationships of N_0, N_1 and N_2 are illustrated in
Fig. 4 as macroscopic dose response curves of amiloride at two
mucosal Na concentrations.

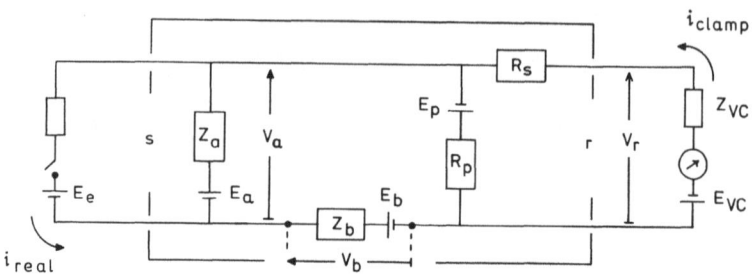

Fig. 5.   The box encloses a two-port filter network made up of
         epithelial elements, solutions and electrodes. At the
         sending port (s), a single channel of the apical
         membrane generates the original noise signal. The
         signal received at the other port (r) by the voltage
         clamp may be attenuated at some or all frequencies. Z_a,
         Z_b, Z_{vc}: impedances of apical and basolateral membrane
         and of the voltage clamp. R_p, R_s: resistances of
         paracellular shunt, solutions and current electrodes
         (Lindemann and DeFelice, 1981).

STEADY STATE NOISE ANALYSIS

     Early results of epithelial noise analysis have been
reviewed previously (Lindemann, 1980). A systematic
investigation of the apical Na-transport began when appropriate
low noise amplifiers for voltage clamping could be constructed
(Van Driessche and Lindemann, 1978), and when it was realized
that amiloride may be used as an extrinsic blocker to generate an
'induced' Lorentzian in the current power density spectrum which
can readily be interpreted (Lindemann and Van Driessche, 1977).

     The structural complexity of epithelia posed some problems.
The presence of a series membrane has two major effects: a)
shifting of the holding potential of the membrane investigated

with respect to the holding potential of the voltage clamp, and
b) attenuation of the noise signal by a filter network which
consists of resistances of membranes and solutions and the
membrane capacitances, as shown in Fig. 5 (Van Driessche and
Gögelein, 1980; Lindemann and DeFelice, 1981).

In principle, the filter effect can be corrected for by
analyzing the tissue impedance and computing the current transfer
function (membrane→clamp) from it. More elegantly, the voltage
transfer function (clamp→membrane) may be estimated, which is
already identical with the required current transfer function
(membrane→clamp) if the experiment is conducted such that the
reciprocity theorem is applicable (Lindemann and DeFelice, 1981).

In the work with apical Na channels effects (a) and (b) were
minimized together by depolarizing the K-permeable basolateral
membrane with a high serosal K concentration. The validity of
this approach is discussed above. In addition the apical
membrane resistance was increased further above the remaining
resistance of the basolateral membrane by the use of high
concentrations of amiloride. Thereby Na_c was kept at low values
such that feedback inhibition was presumably inoperative.

The spontaneous current noise arising in epithelia like frog
skin and toad urinary bladder is considerably enhanced when Na is
present in the mucosal solution. The resulting power density
spectrum shows dominance of low frequency noise essentially
proportional to $1/f^2$ (Fig. 9A). It is conceivable that this
noise component results from spontaneous open/close switching of
Na channels due to the self-inhibition phenomenon.
Unfortunately, the corresponding Lorentzian plateau at very low
frequencies has never been convincingly demonstrated under
voltage clamp conditions, although the spectra have a tendency to
level off below 0.5 Hz. Because of severe stability problems at

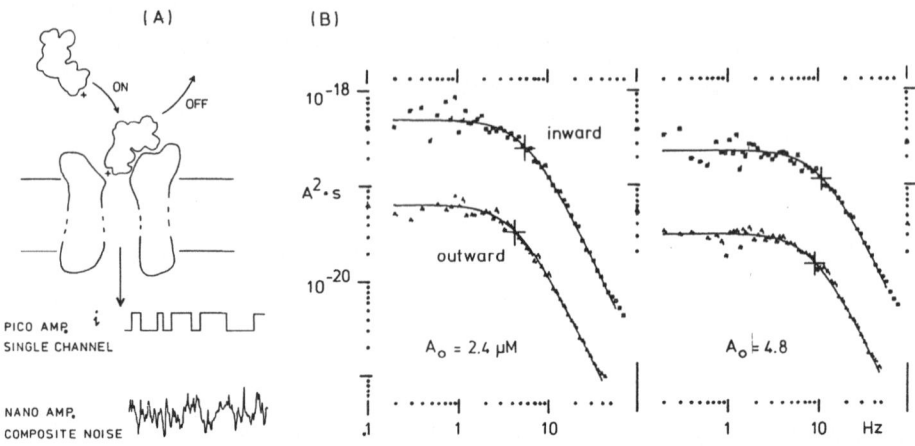

Fig. 6.   (A) Scheme showing how a Na channel might be reversibly
          blocked by amiloride present in the mucosal solution.
          As an alternative to this direct block model (Cuthbert,
          1976), amiloride might bind to a receptor site outside
          of the diffusion pathway and trigger a conformational
          change which closes the channel. Below, the single
          channel current and the composite noise generated by
          additive currents from millions of asynchronous blocking
          events of parallel channels are indicated. (B) Power
          spectra computed from composite noise of K-depolarized
          frog skin, showing amiloride-induced Lorentzians for
          inward (upper curves) and outward flow of Li-ions. The
          solid curves are fits with the Lorentzian function, the
          crosses indicate half plateau powers and corner
          frequencies (Li and Lindemann, 1982).

these low frequencies it may be best to study low frequency
kinetics with other techniques. However, the Lorentzian plateau
was found in voltage noise spectra (Van Driessche and Borghgraef,

1975), and the kinetics of the self-inhibition, as studied with fast concentration steps, point to corner frequencies below 0.2 Hz.

When amiloride is added in maximal concentrations, the Na_o-dependent noise disappears. However, when amiloride is added in submaximal concentrations, the Na_o-dependent noise diminishes at low frequencies while a new Lorentzian component appears in the 1 to 50 Hz band (e.g., Lindemann and Van Driessche, 1977, 1978; Li et al., 1982). The plateaus of the 'amiloride induced' Lorentzians (Fig. 6B and 9B) decrease, and the corner frequencies increase when the amiloride concentration (A_o) is increased. The dependence of corner frequencies (f_c^A) on A_o is linear. Using pseudo-first order kinetics

$$2\pi f_c^A = k_{on} \cdot A_o + k_{off} = 1/\tau_2 \tag{2}$$

the apparent rate constants of amiloride blockage can be obtained from 'rate concentration plots' (Fig. 8 and 9C): k_{on} as the slope, k_{off} as the ordinate intercept. Their ratio

$$K^{mi} = k_{off}/k_{on}$$

will be called the microscopic inhibition constant.

The amiloride-induced Lorentzians were analyzed in terms of competition between blockage by Na_o and A_o (Lindemann and Van Driessche, 1978). In the case of competition with amiloride being the higher rate blocker, we expect

$$k_{on} = k_{02}, \qquad k_{off} = k_{01} \cdot Na_o + k_{20} \tag{3}$$

and therefore

$$K_A^{mi} = K_A(1 + k_{01} \cdot Na_o/k_{20}) = K_A \cdot k_{off}/k_{20} \qquad (4)$$

Indeed, with K-depolarized frog skin f_c^A was observed to increase somewhat with Na_o (Van Driessche and Lindemann, 1979). The effect was small, only 1 Hz (a 10% change) for a tenfold change of Na_o. Since k_{on} was invariant to Na_o, this would imply that k_{off} is at 60 mM Na_o larger than k_{20} by an additive term of about $6 s^{-1}$ (eq. 3). This would be a significant addition since k_{20} is small. However, since it is difficult to resolve a 10% change in f_c^A, we shall just state cautiously that k_{off} may be somewhat larger than k_{20}. Indeed, Hoshiko and Van Driessche (1981) recently found that k_{off} of triamterene, which is larger and therefore more easily measured, did not change significantly with Na_o.

In the same study, Hoshiko and Van Driessche noted that the apparent k_{on} of triamterene decreases when Na_o is raised. These experiments were performed with non-depolarized skins and may mean that Na ions act as higher rate competitors to triamterene, if a membrane voltage which increases the channel current, i, is applied (Machlup and Hoshiko, 1982). However, in K-depolarized skins of near zero apical voltage, k_{on} (of amiloride) did not significantly change with Na_o (Van Driessche and Lindemann, 1979).

The analysis of plateau powers proceeded with the assumption that amiloride blocks the Na channels in a one to one and all or none manner, as illustrated in Fig. 6A. This assumption seems very reasonable since amiloride blocks cellular Na uptake almost completely (e.g., Rick et al., 1975) and in the epithelia used, the Hill coefficient is near unity (e.g., Li et al., 1982). The estimation of the Na channel current, i, and the channel densities N_0, N_1 and N_2 from the plateau powers has been described in detail (Lindemann and Van Driessche, 1978; Li et al., 1982) and need not be repeated in full length. Suffice it to say that competition theory implies a double Lorentzian with two distinct plateaus if the chemical rates of the two blockers are sufficiently far apart. Indeed, the higher frequency Lorentzian usually shows a distinct plateau suitable for analysis if A_o is not too small or too large.

For simple open/close kinetics without competition (K_N very large) the power of the amiloride induced plateau of one-sided spectra (per area) is given by

$$G_o = 4 \, Ni^2 P_0 \, P_2 \tau_2 \tag{5}$$

with

$$I = iNP_0 = iN_o \tag{6}$$

in which I is the macroscopic amiloride blockable Na current per area one obtains

$$\frac{G_o}{I} = 4 \, iP_2\tau_2 = 4 \, i\tau_2 \; \frac{A_o/K_A}{1 + A_o/K_A} \tag{7}$$

Since τ_2 and K_A are obtained from corner frequencies, i can be computed. With competition (K_N small) one obtains the different expression

$$\frac{G_o}{I} = 4 \, i\tau_2 \cdot \frac{P_2}{P_0 + P_2} \tag{8}$$

It may be noted, however, that

$$\frac{P_2}{P_0 + P_2} = \frac{A_o/K_A}{1 + A_o/K_A} \tag{9}$$

which makes eq. 8 identical with eq. 7. However, in case of competition K_A^{mi} rather than K_A is obtained from the rate concentration plot and its use in place of K_A in eq. 8 and eq. 9 must be justified. For $K_A^{mi} = K_A$, we may state that the estimation of i from the Lorentzian plateau of the higher rate blocker will yield the same value with competition kinetics used or neglected. N_o is obtained as the current ratio

$$N_o = I/i \qquad\qquad (4a)$$

provided the microscopic blocking event is 'all or none' (see above). Furthermore, since $P_o + P_2 = P_o \cdot (1 + A_o/K_A)$, the density of directly 'amiloride accessible' channels

$$N_o + N_2 = (I/i) \cdot (1 + A_o/K_A) \qquad\qquad (10)$$

can be calculated. The following, final step of the analysis requires competition theory. In case of competition, the channel density, N_1, is given by

$$N_1 = N \cdot (Na_o/K_N)/(1 + Na_o/K_N + A_o/K_A) \qquad\qquad (11)$$

and should vanish at saturating A_o. Then N can be estimated as $N_o + N_2$ by extrapolating this quantity (eq. 10) to infinite A_o (Li et al., 1982). Thus values for N_o, N_1 and N_2 are obtained.

At near zero membrane voltage i was found to be a linear function of Na_o (Van Driessche and Lindemann, 1979), measured up to 60 mM Na_o. With this activity, more than 10^6 Na ions were calculated to pass an open channel per second. This high translocation rate appeared to be compatible with a channel rather than a carrier of the shuttle type (e.g., Armstrong, 1975). Recent measurements of the flux ratio exponent at 10 and 40 mM Na_o showed the exponent to be very close to unity (Palmer,

1982a; also Dale Benos, personal communication). This very important result excludes an amiloride blockable carrier mechanism and supports the earlier tentative conclusion drawn from the high translocation rate, i.e., that the "Na channels" are indeed channels. With a flux ratio exponent of unity, the channel may have a single-site structure or a multi-site structure of low occupancy. The selectivity series was determined to be H>Li>Na>>K (Palmer, 1982b). The fixed charge density at the outer channel surface appears to be small and to have little effect on the Na translocation rate (Benos et al., 1981).

The estimation of blocking rate constants and channel currents is typically done with Na net flow directed from the mucosal solution into the cells. We attempted to obtain these parameters also under net outward flow of Na ions, using ouabain-blocked frog skins which were preloaded with Na. However, fluctuations and net current into an outer solution of low Na_o were very small under these conditions, presumably because the feedback inhibition had lowered P_{Na} severely. Less inhibition was found when Li rather than Na was used. Amiloride induced Lorentzians were obtained during Li outward flow (Fig. 6B). The blocking rate constants and K_A^{mi} differed a little from the values obtained with Na or Li inward flow: k_{on} was 20% smaller and k_{off} was 47% larger for net Li outward rather than inward transport, both in the presence of ouabain. The computed channel currents appeared smaller, as expected from an increase of the basolateral membrane resistance due to replacement of cellular K by Li ions (Li and Lindemann, 1982).

Once i has been estimated, N_o can be computed and N obtained by extrapolation (eq. 4a and 10). In the toad urinary bladder, a value of 0.5 to 1 channel per μm^2 (per femto-Farad of membrane capacitance) was found (Li et al., 1982). In frog skin, higher

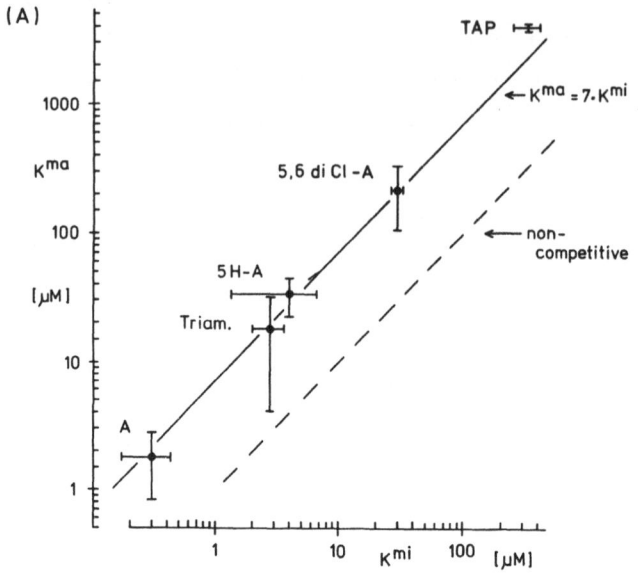

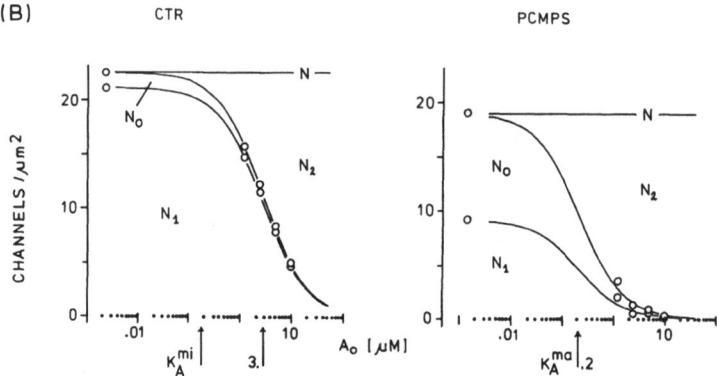

Fig. 7. (A) Relationship between macroscopic (K^{ma}) and
microscopic (K^{mi}) inhibition constants as obtained for
amiloride, triamterene, 5 H-amiloride, 5,6
diCl-amiloride and triaminopyrimidine. K-depolarized
frog skin. The bars indicate one standard deviation.
The dashed line is valid for non-competitive inhibition
($K^{ma} = K^{mi}$). The fact that K^{ma} is in the mean sevenfold
larger than K^{mi} (which was determined from rate
concentration plots as k_{off}/k_{on}) indicates a strong

densities of around 30 channels per μm^2 or about 2000 per granular cell were obtained for the presence of 60 mM Na_o. N_o increased when Na_o was lowered, in rough agreement with the expected competition behavior (Van Driessche and Lindemann, 1979). Furthermore, the sum $N_o + N_2$ increased with A_o (Li et al., 1982). It must be emphasized that these steady state noise experiments do not show by themselves whether the decrease of N_o with increasing Na_o is due to self-inhibition or feedback inhibition. However, with the pump running, Na_c was estimated to be no larger than 6 mM at 60 mM Na_o (Fuchs et al., 1977 and unpublished). In the presence of amiloride, the value should be even smaller. Therefore, Na_c-dependent feedback inhibition is unlikely to contribute much to the observed variation in N_o.

More recently, the inhibition constants from macroscopic dose response curves and those obtained from corner frequencies (i.e., 'microscopically') were compared for 5 extrinsic blockers including amiloride (Li and Lindemann, 1983a). It was found with 60 mM Na_o that

$$K_A{}^{ma} = K_A(1 + Na_o/K_N) \tag{12}$$

was in the mean seven-fold larger than $K_A{}^{mi}$ (Fig. 7A). In conjunction with eq. 4, we obtain the theoretical relationship

competitive component in the blocking mechanism of the five compounds (Li and Lindemann, 1983a). (B) Effect of 0.1 mM PCMPS on channel densities, plotted as a function of the amiloride concentration. The decrease of $K_A{}^{ma}$ with respect to the control (left) and the increase in N_o at the expense of N_1 are evident. K-depolarized frog skin (Li and Lindemann, 1982b).

$$\frac{K_A^{ma}}{K_A^{mi}} = \frac{1 + Na_o/K_N}{k_{off}/k_{20}} \qquad (13)$$

which shows that the result of Fig. 7A is expected from
competition kinetics if $K_N \sim 10$ mM and $k_{off} \sim k_{20}$. For $k_{off} > k_{20}$,
K_N should be even smaller. These data are also compatible with
mixed inhibition rather than pure competition, but only if the
degree of competition exceeds the non-competitive part at least
seven-fold.

As mentioned above, macroscopic experiments have indicated
that agents like BIG and PCMB and PCMPS release the Na channels
from the self-inhibition. Results obtained by noise analysis in
frog skin are in accord with this conclusion. The agents
increased N_o at the expense of N_1 while K_A^{ma} decreased and i was
unaffected or only marginally changed (Li and Lindemann, 1983b).
The total channel density N was somewhat decreased by BIG but
less affected by PCMPS (Fig. 7B). In addition, BIG was noted to
be a high rate blocking competitor of amiloride, as previously
concluded from macroscopic data (Zeiske and Lindemann, 1974).

In the work described above extrinsic blockers were used
extensively as tools. It is desirable, therefore, to investigate
their mode of action more closely. This was attempted in a
structure-activity study in which the blocking rate constants of
25 structural analogues of amiloride were obtained (Li et al.,
1984). It was found that the substituents at position 5 and 6 of
the pyrazine ring essentially control the off-rate and,
therefore, the duration of the blocking event. The on-rate
constant is not affected by halo-substitutions at position 6
(Fig. 8A,B). The on-rate constant is typically much smaller than
expected from diffusion-limited encounters. It is not
concentration dependent but differs widely among analogues (e.g.,

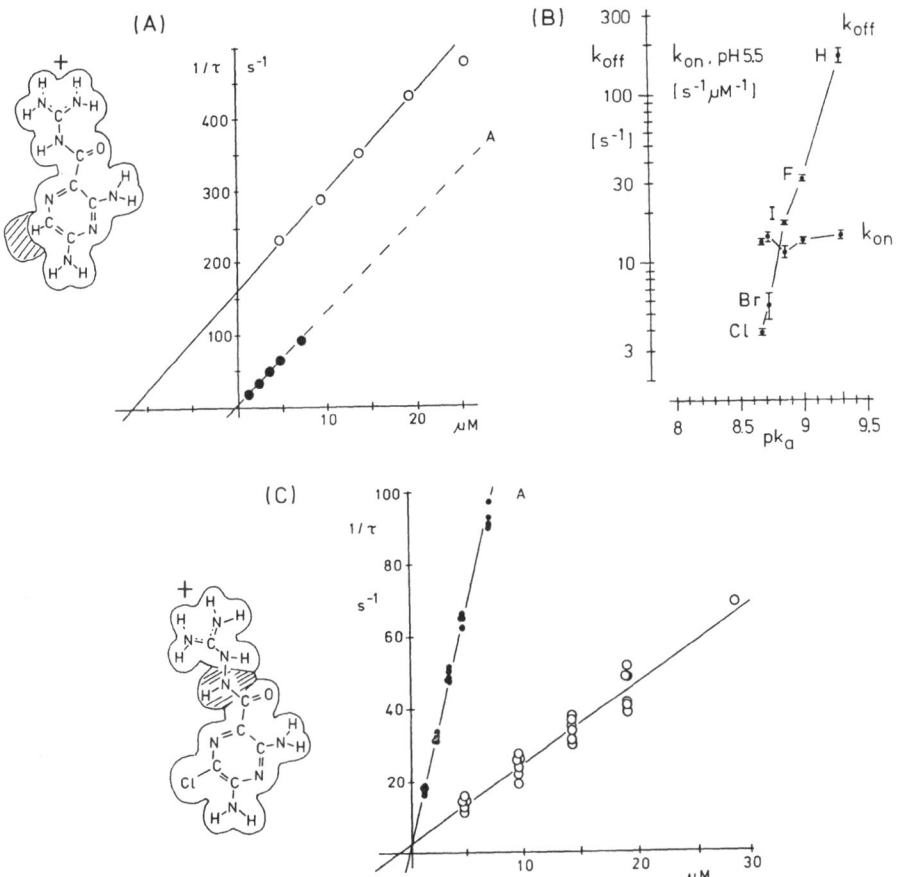

Fig. 8.   (A) Rate concentration plot for 6H-amiloride.   By
substituting the chlorine at position 6 by hydrogen,
k_{off} increases with respect to amiloride (dashed line)
while k_{on} (slope) is not affected.   (B) The blocking
rate constants are plotted against the pK_a of the
amidino group for halo substitutions at position 6.   The
bars indicate one standard error of the mean.   (C) Rate
concentration plot for an analogue with prolonged 'side
chain'.   The on-rate constant (slope) is lowered with
respect to that of amiloride (filled circles).
K-depolarized frog skin. (Li et al., 1983).

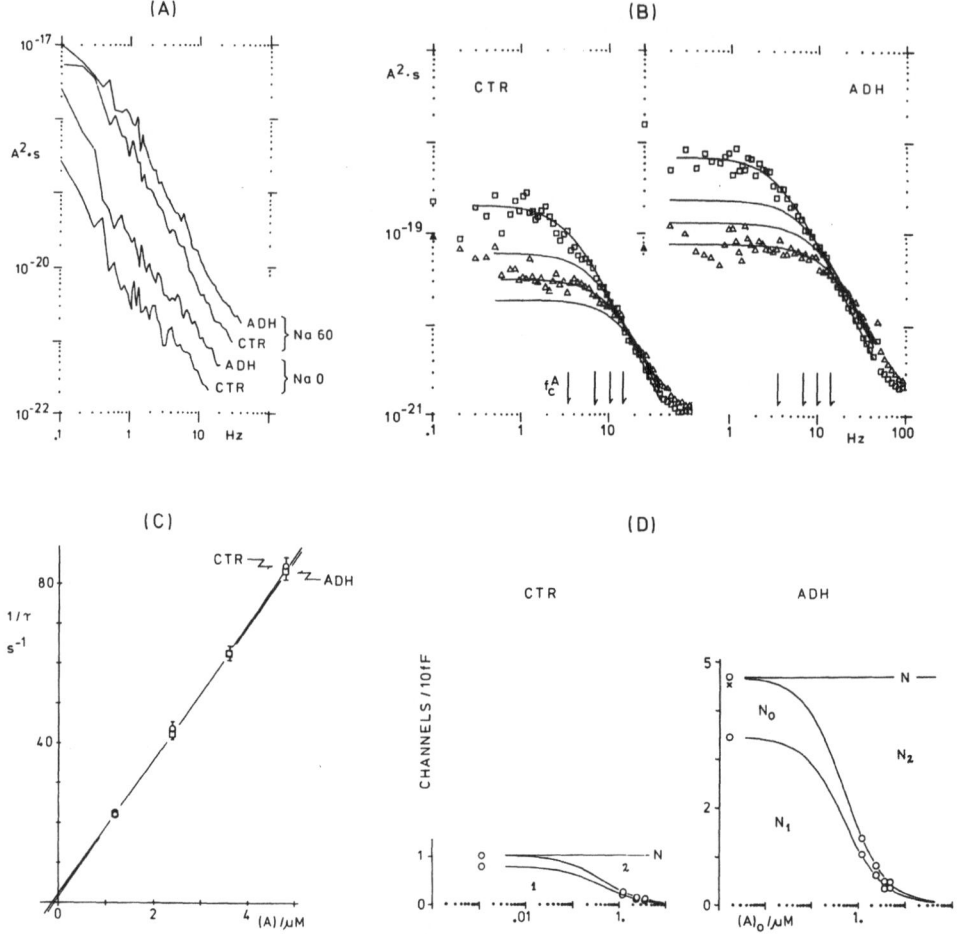

Fig. 9.   Noise analysis of ADH effects in toad urinary bladder.
(A) Power density spectra calculated from spontaneous
noise (in the absence of amiloride). CTR=control. (B)
Lorentzians induced by amiloride in four concentrations
(see panel C) before and during ADH-stimulation.   (C)
Rate concentration plot. The bars denote one standard
error of the mean. (D) Calculated channel densities
plotted against the amiloride concentration before (CTR)
and during ADH stimulation (Li et al., 1982).

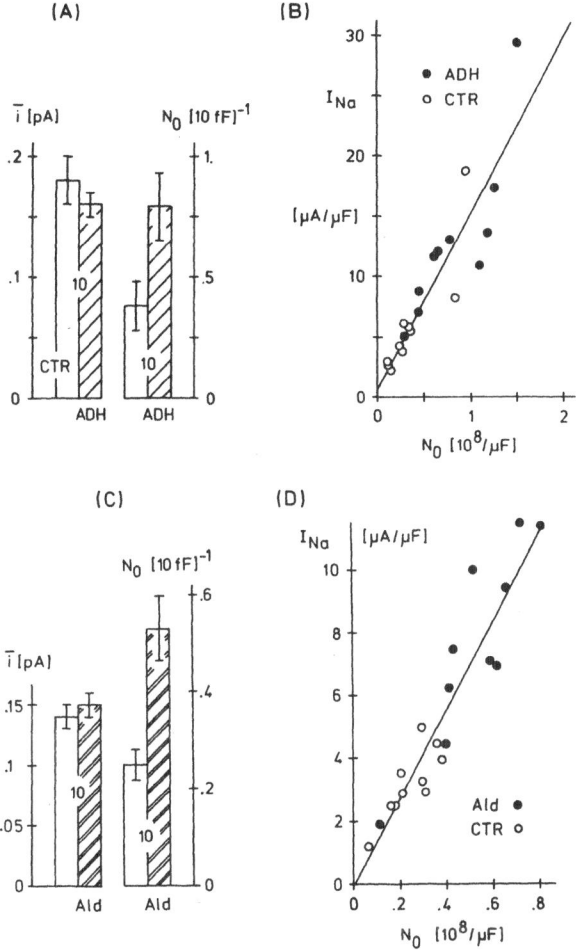

Fig. 10. Effects of ADH (panel A, B) and aldosterone (panel C,D)
on i and N_o in toad urinary bladder (Li et al., 1982;
Palmer et al., 1982).

Fig. 8C). This 'structure-dependence' of the on-rate is not
convincingly explained by steric hindrance. It was concluded,
therefore, that the structure-dependence of k_{on} results from
interactions probably of side-chain ligands with the channel

prior to the actual blocking. Analysis of the corresponding
kinetic equations showed that the lifetime of the 'encounter
complex' must be short if the rate concentration plots are
linear, as they are. It is interesting that the formation of an
encounter complex provides a possibility to explain the
stimulatory effect of some amiloride analogues, if it is
postulated that formation of this complex is competitive to the
self-inhibition and that the channel conducts for the lifetime of
this complex.

Hormonal Effects on Na-channels

Once i and N_o could be estimated by noise analysis it became
feasible to study the effects of Na-transport stimulating
hormones in terms of these parameters. The antidiuretic hormone
(ADH) is known to initially increase the cellular concentration
of cyclic AMP (e.g., Orloff and Handler, 1967). The resulting
stimulation of protein kinase then leads through a number of
unknown steps, which possibly involve dephosphorylation of
membrane proteins (e.g., Walton et al., 1975), to an increase in
P_{Na}. Noise analysis (Li et al, 1979, 1982; Helman et al., 1981)
showed that an increase in channel currents is not involved in
this process. Rather, N_o increases, however, not at the expense
of N_1. Instead, the total number of channels, N, is found to be
increased (Fig. 9,10) in proportion to the increase in P_{Na}. It
appears, therefore, that by this hormone more Na channels are
made available for transport and for regulatory inhibition. We
have called this increase in N recruitment.

Since ADH is a fast hormone which stimulates apical
Na-transport within 60 to 120 seconds, one would perhaps expect
that there is an electrically silent population of Na channels in
the apical membrane, which can be rapidly recruited by a small
chemical change, for instance by dephosphorylation. However,

recent experiments by Garty and Edelman (1983) have shown that prior to recruitment these channels are not available to attack by trypsin present in the mucosal solution. It seems more likely, therefore, that the channels recruited by ADH- or cyclic AMP are contained in the walls of sub-apical vesicles to which trypsin has no access, and that ADH stimulated recruitment involves fusion of these vesicles with the membrane.

Fusion of submembranous vesicles into the apical membrane should increase the electrical membrane capacitance. In toad urinary bladder, a surprisingly large increase in capacitance was actually found (Warncke and Lindemann, 1980; Stetson et al., 1981). It occurs in the absence of osmotic gradients, i.e., under conditions where gross morphological changes of the apical membrane (e.g., Davis et al., 1974) caused by ADH should be limited (Spinelli et al, 1975). Stetson et al. (1981) found that methohexital blocks the capacitance increase, which therefore may be due entirely to the fusion events of the parallel hydrosmotic effect of ADH (Wade et al., 1981). However, in our hands, methohexital had side-effects on Na transport (Warncke and Lindemann, unpublished). While the interpretation of the capacitance increase is perhaps not yet entirely clear, the trypsin data are compatible with an ADH-dependent vesicular recruitment of Na channels.

Aldosterone stimulates Na-transport after an induction time of several hours. It induces several proteins, of which one has been identified, citrate synthase (Law and Edelman, 1978). It is unknown whether the increase in P_{Na} observed in toad urinary bladder in response to aldosterone application is in part due to de novo synthesis of Na channels. Noise analysis showed that the hormone increases both N_o and N (Fig. 10C,D) while leaving i essentially unchanged (Palmer et al., 1982). In this respect the response is similar to the effect of ADH, although much slower.

The long response time may be due to the fact that metabolic
enzymes like citrate synthase have to be synthesized, which –
possibly through products of energy metabolism – permit
recruitment from an electrically silent pool of Na channels.
Indeed, when energy metabolism is impaired, P_{Na} decreases
considerably, presumably because N is decreased (Garty et al.,
1983), and during mild metabolic impairment aldosterone is unable
to increase N although the induction takes place. In short,
aldosterone causes recruitment, but this effect depends on and
perhaps is mediated by 'energy metabolism'.

Clearly the reservoir of channels from which aldosterone
recruits need not be identical with that from which cyclic AMP
recruits. Indeed, Garty and Edelman (1984) found that mucosal
trypsin attacks the channels which are later recruited by
aldosterone. Amiloride protects from the tryptic split.
Therefore, the reservoirs for ADH and aldosterone stimulated
recruitment apparently are at different locations, that for
aldosterone recruitment perhaps within the apical membrane.
Alternatively, this reservoir may also be vesicular, provided it
allows for access of trypsin. This might happen by the
fusion/defusion turnover being faster than in the case of
ADH-recruited vesicles. For the mammalian urinary bladder, Lewis
and deMoura (1982) found that the subapical vesicles contain Na
channels of a density which is increased by aldosterone, and that
these vesicles can be made to fuse with the apical membrane.

The emerging, as yet certainly incomplete picture of
inhibitory and recruiting processes has been represented as Fig.
11. An additional regulatory mechanism, recruitment of
basolateral pump-leak units, has recently been suggested (Thomas
et al., 1983).

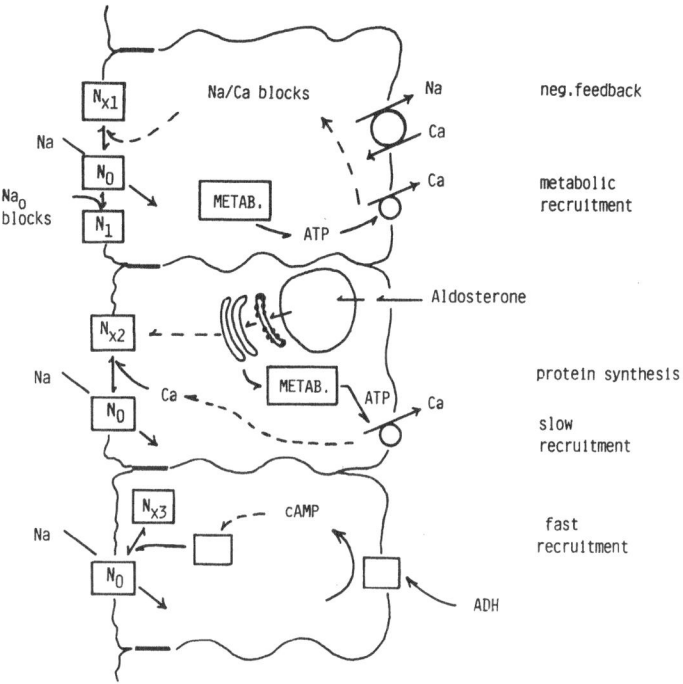

Fig. 11. Modulation of Na channel density by various control
         mechanisms. The dashed lines depict hypothetical steps,
         showing how much is presently uncertain. Upper cell:
         self-inhibition is represented as a 'fast turnoff'
         mechanism on the left. Negative feedback is shown to
         depend on cytosolic Ca, modulated by basolateral Na/Ca
         exchange and an ATP driven Ca pump. It involves channel
         reservoir N_{X1}. In addition, the necessity of active Ca
         extrusion or sequestration may explain the dependence of
         P_{Na} on the rate of energy metabolism (metabolic
         recruitment). The middle cell depicts slow recruitment
         by aldosterone from reservoir N_{X2}, involving channel
         synthesis (?) and an unknown process depending on energy
         metabolism. It is unknown whether N_{X1} and N_{X2} are
         different reservoirs. The lower cell depicts fast
         ADH-dependent recruitment. N_{X3} is very likely vesicular
         and different from N_{X2}. For more details see text.

CONCLUSION

In summary, then, the apical Na translocators appear to be
channels of high Na,Li selectivity which permit large transport
rates at low driving forces. The concentration dependence of the
transport rate is linear up to 60 mM Na$_o$. The voltage-dependence
is of the Goldman-Hodgkin-Katz type (only macroscopic results).
Outward facing surface charges have little influence. The flux
ratio exponent is close to unity, compatible with single site
channels or multi-site channels of low occupancy. The cellular
and hormonal regulatory mechanisms vary Na-transport by changing
the density of conducting channels in a variety of ways.

ACKNOWLEDGEMENT

We are grateful to Frau U. Lang and to Frau B. Hasper for
processing text and figures. Frau C. Bauer and Dr. T. D. Plant
kindly improved the English. Support was obtained from the
Deutsche Forschungsgemeinschaft through SFB 38, project C1.

REFERENCES

Aceves, J., and Cuthbert, A. W., 1979, Uptake of [3H] Benzamil at
    different sodium concentrations. Inferences regarding the
    regulation of sodium permeability, J. Physiol., 295:491.
Armstrong, C. M., 1975, Evidence for ionic pores in excitable
    membranes, Biophysical J., 15:932.
Benos, D., Latorre, R., and Reyes, J., 1981, Surface potentials
    and sodium entry in frog skin epithelium, J. Physiol.,
    321:163.

Bevevino, L. H., and Lacaz-Vieira, F., 1982, Control of sodium
    permeability of the outer barrier in toad skin, J. Membrane
    Biol., 66:97.

Bindslev, N., Cuthbert, A. W., Edwardson, J. M., and Skadhauge,
    E., 1982, Kinetics of amiloride action in the hen coprodaeum
    in vitro, Pflügers Arch., 392:340.

Cereijido, M., Herrera, F. C., Flanigan, W. J., and Curran, P.
    F., 1964, The influence of Na concentration on Na transport
    across frog skin, J. Gen. Physiol., 47:879.

Chase, H. S., and Al-Awqati, Q., 1981, Regulation of the sodium
    permeability of the luminal border of toad bladder by
    intracellular sodium and calcium. Role of sodium-calcium
    exchange in the basolateral membrane, J. Gen. Physiol.,
    77:693.

Chase, H. S., and Al-Awqati, Q., 1982, Submicromolar calcium
    regulates Na permeability of luminal membrane vesicles from
    toad bladder as measured by flow quench method, Fed. Proc.
    41:1350.

Cuthbert, A. W., 1976, Importance of guanidinium groups for
    blocking sodium channels in epithelia, Mol. Pharmacol.,
    12:945.

Cuthbert, A. W., and Shum, W. K., 1974, Binding of amiloride to
    sodium channels in frog skin, Mol. Pharmacol., 10:880.

Davis, W. L., Goodman, D. B. P., Martin, J. H., Mathews, J. L.,
    and Rasmussen, H., 1974, Vasopressin-induced changes in the
    toad urinary bladder epithelial surface, J. Cell Biol.,
    61:544.

Dick, H. J., and Lindemann, B., 1975, Saturation of Na-current
    into frog skin epithelium abolished by PCMPS, Pflügers
    Arch., 355:R72.

Fisher, R. S., and Helman, S. I., 1981, Influence of basolateral
    $(K)_i$ on the electrical parameters of the cells of isolated
    epithelia of frog skin, Biophysical J., 33:41a.

Frömter, E., Higgins, J. T., and Gebler, B., 1981, Electrical
    properties of amphibian urinary bladder epithelia. IV. The
    current-voltage relationship of the sodium channels in the
    apical cell membrane, in: "Ion Transport by Epithelia,".
    S. G. Schultz, ed., Raven Press, New York.

Fuchs, W., Hviid Larsen, E., and Lindemann, B., 1977,
    Current-voltage curve of sodium channels and concentration
    dependence of sodium permeability in frog skin, J. Physiol.,
    267:137.

Garty, H., and Edelman, I. S., 1983, Amiloride-sensitive
    trypsinization of apical sodium channels: Analysis of
    hormonal regulation of sodium transport in toad bladder, J.
    Gen. Physiol., 81:785.

Garty, H., Edelman, I. S., and Lindemann, B., 1983, Metabolic
    regulation of apical sodium permeability in toad bladder in
    the presence and absence of aldosterone, J. Membrane Biol.,
    74:15.

Garty, H. and Lindemann, B., 1984, Feedback inhibition of sodium
    uptake in K-depolarized toad urinary bladders, J. Membrane
    Biol., submitted.

Grinstein, S., and Erlij, D., 1978, Intracellular calcium and the
    regulation of sodium transport in the frog skin, Proc. R.
    Soc. Lond. B. 202:353.

Harms, V., and Fanestil, D. D., 1977, Functions of apical
    membrane of toad urinary bladder: effects of membrane
    impermeant reagents, Am. J. Physiol., 233:F607.

Helman, S. I., Cox, T. C., and Van Driessche, W., 1981, Changes
    of Na channel number at the apical membrane of frog skin
    caused by indomethacine and ADH/theophylline, Abstracts VII
    International Biophysics Congress, Mexico, p. 184.

Hoshiko, T., and Van Driessche, W., 1981, Triamterene-induced
    sodium current fluctuations in frog skin, Arch. Int.
    Physiol. Biochem., 89:58.

Koefoed-Johnsen, V., and Ussing, H. H., 1958, The nature of the frog skin potential, Acta Physiol. Scand., 42:298.

Law, P. Y., and Edelman, I. S., 1978, Induction of citrate synthase by aldosterone in the rat kidney, J. Membrane Biol., 41:41.

Leblanc, G., and Morel, F., 1975, Na and K movements across the membranes of frog skin epithelia associated with transient current changes, Pflügers Arch., 358:159.

Lewis, S. A. and de Moura, J. L. C., 1982, Incorporation of cytoplasmic vesicles into apical membrane of mammalian urinary bladder epithelium, Nature, 197:685.

Li, J. H.-Y., and Lindemann, B., 1982, Movement of Na and Li across the apical membrane of frog skin, in: "Basic Mechanisms in the Action of Lithium,". H. M. Emrich, J. B. Aldenhoff and H. D. Lux, ed., Excerpta Medica, Amsterdam.

Li, J. H.-Y., and Lindemann, B., 1983a, Competitive blocking of epithelial Na channels by organic cations: the relationship between macroscopic and microscopic inhibition constants, J. Membrane Biol.,76:235.

Li, J. H.-Y., and Lindemann, B., 1983b, Chemical stimulation of Na Transport through amiloride-blockable channels of frog skin epithelium, J. Membrane Biol.,75:179.

Li, J. H.-Y., Cragoe, Jr., E. J., and Lindemann, B., 1984, Structure activity relationship of amiloride analogues, J. Membrane Biol., submitted.

Li, J. H.-Y, Palmer, L.G., Edelman, I. S. and Lindemann, B., 1979, Effect of ADH on Na channel parameters in toad urinary bladder, Pflügers Arch., 382:R13.

Li, J. H.-Y., Palmer, L. G., Edelman, I. S., Lindemann, B., 1982. The role of Na-channel density in the natriferic response of the toad urinary bladder to an antidiuretic hormone, J. Membrane Biol., 64:77.

Lindemann, B., and Gebhardt, U., 1973, Delayed changes of
    Na-permeability in response to steps of $(Na)_o$ at the outer
    surface of frog skin and toad bladder, in: "Transport
    mechanisms in epithelia," H. H. Ussing and N. A. Thorn,
    Munksgaard, Copenhagen.

Lindemann, B., 1980, The beginning of fluctuation analysis of
    epithelial ion transport, J. Membrane Biol., 54:1.

Lindemann, B., 1977, A modifier-site model for passive Na
    transport into frog skin epithelium, in: "Intestinal
    Permeation," M. Kramer and F. Lauterbach, ed., Excerpta
    Medica, Amsterdam.

Lindemann, B., and DeFelice, L. J., 1981, On the use of general
    network functions in the evaluation of noise spectra
    obtained from epithelia, in: "Ion Transport by Epithelia,"
    S.G. Schultz, ed., Raven Press, New York.

Lindemann, B., and Van Driessche, W., 1977, Sodium specific
    membrane channels of frog skin are pores: current
    fluctuations reveal high turnover, Science, 195:292.

Lindemann, B., and Van Driessche, W., 1978, The mechanism of Na
    uptake through Na-selective channels in the epithelium of
    frog skin, in: "Membrane Transport Processes," Vol. 1, J. F.
    Hoffman, ed., Raven Press, New York.

Machlup, S., and Hoshiko, T., 1982, Sodium and amiloride
    competition in apical membrane channels: a 3-state model for
    noise, Biophysical J., 37:281a.

MacRobbie, E. A. C., and Ussing, H. H., 1961, Osmotic behavior of
    the epithelial cells of frog skin, Acta Physiol. Scand.,
    53:348.

Morel, F., and Leblanc, G., 1975, Transient current changes and
    Na compartmentalization in frog skin epithelium, Pflügers
    Arch., 358:135.

Nelson, D. J., Ehrenfeld, J., and Lindemann, B., 1978, Volume
    changes and potential artifacts of epithelial cells of frog
    skin following impalement with microelectrodes filled with 3
    M KCl, J. Membrane Biol., 40:91.

Orloff, J., and Handler, J., 1967, The role of adenosine 3, '5'
    -phosphate in the action of antidiuretic hormone, Am. J.
    Med., 42:757.

Palmer, L. G., 1982a, Na$^+$ transport and flux ratio through apical
    Na$^+$ channels in toad bladder, Nature, 297:688.

Palmer, L. G., 1982b, Ion selectivity of the apical membrane Na
    channel in the toad urinary bladder, J. Membrane Biol.,
    67:91.

Palmer, L. G., Edelman, I. S., and Lindemann, B., 1981,
    Current-voltage analysis of apical Na transport in toad
    urinary bladder: effects of inhibitors of transport and
    metabolism, J. Membrane Biol., 57:59.

Rawlins, F., Mateu, L., Fragachan, F., and Whittembury, G., 1970,
    Isolated toad skin epithelium: transport characteristics,
    Pflügers Arch., 316:64.

Rick, R., Dorge, A., and Nagel, W., 1975, Influx and efflux of
    sodium at the outer surface of frog skin, J. Membrane Biol.,
    22:183.

Spinelli, F., Gross, A., and de Sousa, R. C., 1975, The
    hydrosmotic effect of vasopressin: a scanning electron
    microscope study, J. Membrane Biol., 23:139.

Stetson, D. L., Lewis, S. A., and Wade, J. B., 1981, ADH-induced
    increase in transepithelial capacitance in toad bladder,
    Biophysical J., 33:43a.

Sudou, K., and Hoshi, T., 1977, Mode of action of amiloride in
    toad urinary bladder. An electrophysiological study of the
    drug action on sodium permeability of the mucosal border, J.
    Membrane Biol., 32:115.

Taylor, A., and Windhager, E. E., 1979, Possible role of
    cytosolic calcium and Na-Ca exchange in regulation of
    transepithelial sodium transport, Am. J. Physiol., 236:F505.

Thomas, S. R., Suzuki, Y., Thompson, S. M., and Schultz, S. G.,
    1983, Electrophysiology of necturus urinary bladder: I.
    'Instantaneous' current-voltage relations in the presence of
    varying mucosal sodium concentrations, J. Membrane Biol.,
    73:157.

Thompson, S. M., Suzuki, Y., and Schultz, S. G., 1982, The
    electrophysiology of rabbit descending colon. I.
    'Instantaneous' transepithelial current-voltage relations
    and the current-voltage relations of the Na-entry mechanism,
    J. Membrane Biol., 66:41.

Ussing, H. H., 1949, The active ion transport through the
    isolated frog skin in the light of tracer studies, Acta
    Physiol. Scand., 17:1.

Van Driessche, W., and Borghgraef, R., 1975, Noise generated
    during ion transport across frog skin, Arch. Int. Physiol.
    Biochim., 83:140.

Van Driessche, W., and Goegelein, H., 1980, Attenuation of
    current and voltage noise signals recorded from epithelia,
    J. Theor. Biol., 86:629.

Van Driessche, W., and Lindemann, B., 1978, Low-noise
    amplification of voltage and current fluctuations arising in
    epithelia, Rev. Sci. Instrum., 49:52.

Van Driessche, W., and Lindemann, B., 1979,
    Concentration-dependence of currents through single
    sodium-selective pores in frog skin, Nature, 282:519.

Wade, J. B., Stetson, D. L., and Lewis, A. L., 1981, ADH action:
    evidence for a membrane shutttle mechanism, Ann. N.Y. Acad.
    Sci., 372:106.

Walton, K. G., DeLorenzo, R. J., Curran, P. F., and Greengard,
    P., 1975, Regulation of protein phosphorylation in sodium
    transport in toad bladder, J. Gen. Physiol., 65:153.

Warncke, J., and Lindemann, B., 1980, Effect of ADH on the capacitance of apical epithelial membranes, Adv. Physiol. Sci., 3:129.

Zeiske, W., and Lindemann, B., 1974, Chemical stimulation of Na current through the outer surface of frog skin epithelium, Biochem. Biophys. Acta, 352:323.

THE INFORMATION CONTENT OF SINGLE CHANNEL DATA

Joseph Patlak

Department of Physiology and Biophysics
University of Vermont
Burlington, VT 05446, USA

INTRODUCTION

Single channel signals were initially observed in artificial
bilayer membranes (Bean et al., 1969; Gordon and Haydon, 1972;
Hladkey and Haydon, 1972). They were first seen in biological
membranes using the patch clamp technique (Neher and Sakmann,
1976). A number of excellent reviews have detailed this
technique, including gigohm ($>10^9$ ohm) seals and ultra-high
resolution recordings, as well as the many different types of
preparations and channels from which recordings have been made
(Neher et al., 1978; Hamill et al., 1981). In this chapter I will
examine a different aspect of this subject. I seek to clarify the
reasons for recording single channels, instead of measuring the
mean behavior of a population of channels (whole cell recording),
or the fluctuations caused by the random activity of such a
population.

The advantages of single channel recording are complex. They
are heavily influenced by the technical advantages of patch clamp
recording, such as separation of current components, control of

bathing solutions on both sides of the membrane, voltage clamping
of small cells, etc. However, recording single channel signals
also provides distinct advantages in the quality and quantity of
information that can be gained. This chapter will discuss these
advantages, and also some of the ways in which single channel
signals can be used to obtain information that would not be
available using other recording techniques.

     The following four advantages of single channel recording
will be discussed: 1) The single channel current, and thereby the
single channel conductance, is measured directly, not calculated
on the basis of a specific model and several assumptions. 2) The
measurement of single channel kinetics is more sensitive than
fluctuation analysis for detecting heterogeneity in channel
populations or in channel kinetics. 3) Single channel recording
provides more independent measurements of a channel's kinetics and
allows individual rate constants in a particular kinetic model to
be independently assessed. 4) The method also permits the
identification of the state of a given channel at a particular
time, so that conditional probability and other statistical
analyses can be made.

SINGLE CHANNEL CONDUCTANCE

     The currents through individual channels can be observed in
patches of membrane in which channels have either a low density or
a low probability of opening. Discrete pulses of current
originating at and returning to the zero current level are
observed. The currents jump quickly to a steady level of flow,
and stay at that level until they quickly return to the zero
current level. This behavior has been reported in almost every

type of channel that has been observed with the single channel
technique. By measuring the difference in the current level
between the baseline and the open channel level one gets a measure
of the amount of current which flows through a channel. If the
trans-membrane potential and the reversal potential for currents
through that channel are known, then the single channel
conductance can be calculated directly using Ohms law,

$$\gamma = I_s / (E_m - E_o)$$

where γ is the single channel conductance, I_s is the current
through the open channel, E_m is the membrane potential, and E_o is
the potential at which the net movement of charge through the
channel is zero (reversal potential). Single channel recording
with excised membrane patches (Horn and Patlak, 1980), or with
artificial bilayer membranes, permits both the trans-membrane
potential and the bathing solutions on both sides of the membrane
to be experimentally controlled. Thus an accurate determination
of both E_m and E_o in the above equation can be obtained.

Fluctuation analysis, however, calculates single channel
conductance by measuring the variance and mean current for a
fluctuating signal (Katz and Miledi, 1970; Anderson and Stevens,
1973). When there is a low probability that any one channel is
open, the single channel current is the ratio of these two
quantities. However, several assumptions are implicit even in
this simple case. First, one assumes that channels have only two
states, open and closed. Second, that they are not bursting or
flickering between states when 'open', and third that only one
type of channel is present and conducting at one time. These
assumptions are usually justified (as evidenced by success of
fluctuation analysis), but seldom can be explicitly demonstrated.
The analysis of single channel signals is advantageous because it
is not necessary to make these assumptions.

MEASURING CHANNEL KINETICS

    Implicit in the concept of membrane channels is the notion
that these channels exist in either conducting or non-conducting
configurations. Since channels are membrane-bound macromolecules
with complex structure, they probably have several different
conformational states which they can assume, and some of these
states are conducting. The switching of the channel between such
conformational states gives rise to the quantum jumps of current
seen during single channel recording. This molecular concept of
channels is especially appealing because much of the kinetic
behavior that they exhibit can be explained by the stochastic
switching of the channel between two or more conformational
states.

    Stochastic models of channel activity can be constructed
using several basic assumptions about random processes (Colquhoun
and Hawkes, 1981). These assumptions are outlined as follows: 1)
Channels exist in discrete 'states' which have lifetimes much
longer than the time needed to switch between states. 2) The
probability that a given transition will occur is dependent only
on the channel's current state, and not on its past history. 3)
The rate of switching between two states is determined by a single
parameter, the rate constant, which is invariant with time, but
may depend on external conditions such as voltage or temperature.
The consequence of these three assumptions is that the lifetime of
any state is described by a single exponential function whose time
constant is the inverse of the sum of all rates leaving that
state. For example, a channel might have three discrete states,
two closed and one open. A state diagram for such a channel could
be as follows:

$$C_1 \underset{k_{-1}}{\overset{k_1}{\rightleftharpoons}} C_2 \underset{k_{-2}}{\overset{k_2}{\rightleftharpoons}} 0$$

where C_1 and C_2 are closed states, and 0 is the open state. According to the concept outlined above, the mean lifetime of the open state is the inverse of the rate for leaving that state, $1/k_{-2}$. The mean lifetime of the intermediate state, C_2, is the inverse of all the rates for leaving the state, $1/(k_2 + k_{-1})$. In the case of this simple channel, the empirical expression for the lifetime in each state is relatively easy to obtain using differential equations or linear algebra techniques.

One goal of membrane channel studies is to identify the possible kinetic states for a channel, and to characterize the rate constants with which channels can change states. The single channel recording technique can be used to measure several unique types of kinetic information about channel behavior. Such analyses provide more information about channel kinetics than any other conventional method, as will be outlined in the following pages.

Channel Open Time

One of the easiest examples of kinetic analysis using single channel recording is the measurement of the channel's open time. According to the stochastic model, each time the channel opens, it is entering a distinct kinetic state. The amount of time it spends in this state is different for each opening, but the average duration of its stay will be a constant, depending only on the recording conditions and the type of channel. The measurement of this open time is straightforward and seldom ambiguous: one simply measures the duration between the channel's entry into its open state, and the first instance of it leaving that state, i.e.,

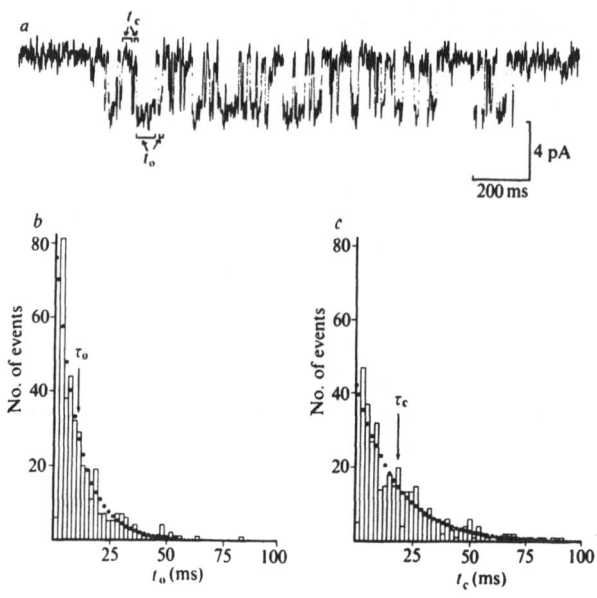

Fig. 1.   Analysis of the rapid fluctuation of current during a
burst from an ACh activated channel on frog cutaneous
pectoris muscle. (A). A single recorded burst in the
presence of 10 μM ACh. Brackets indicate the measurement
of individual channel open times, τ_o and closed times τ_c.
The membrane potential was -130 mV, and the temperature
12°C. Note the high level of noise in this recording
made before the discovery of giga-ohm seals.
(B) and (C) are histograms of distributions of open and
closed times measured during several sequential bursts,
including the one shown in (A). The histograms were
fitted with single exponential curves, as indicated by
the dotted line, using a least-squares fitting routine
weighted for the statistical nature of the underlying
events. The arrows indicate the best fit decay constants
τ_o and τ_c for the distribution of open and closed times,
respectively. τ_o was 10.6 ms and τ_c was 18 ms. (From
Sakmann et al., 1980).

closing. Such measurements can be made for many events, and the values thus obtained may be averaged to give a final value.

This sort of kinetic measurement is actually a measure of the (conditional) probability--given that the channel entered the open state at time zero--that the channel is still in the open state for all subsequent times. For a channel with a single open state that satisfies the definitions given above, this conditional probability will be expressed as a probability density function with an initial value of one, and with a time constant for the exponential decay equal to the inverse of the rate for leaving the open state.

As described above, such a conditional probability measurement can be made by constructing a histogram of measured open times for many individual events. An example of this type of measurement is shown in Fig. 1. Open times were measured for extra-synaptic acetylcholine activated channels in frog muscle during bursts of activity where only one channel was active at any time. A histogram of the channel open times is shown in Fig. 1B. This histogram is well fitted by a single exponential, and is consistent with the idea that this channel has a single open state that fits the definitions listed above.

In contrast, fluctuation analysis and macroscopic recording cannot measure just those events which fulfill a particular condition, but instead must analyze all events together. Only the sum of all rates together can be measured. Although the experimenter can often devise conditions where one rate dominates the measured data, thereby obtaining meaningful information from fluctuation and whole cell recordings, single channel data often avoid such complications.

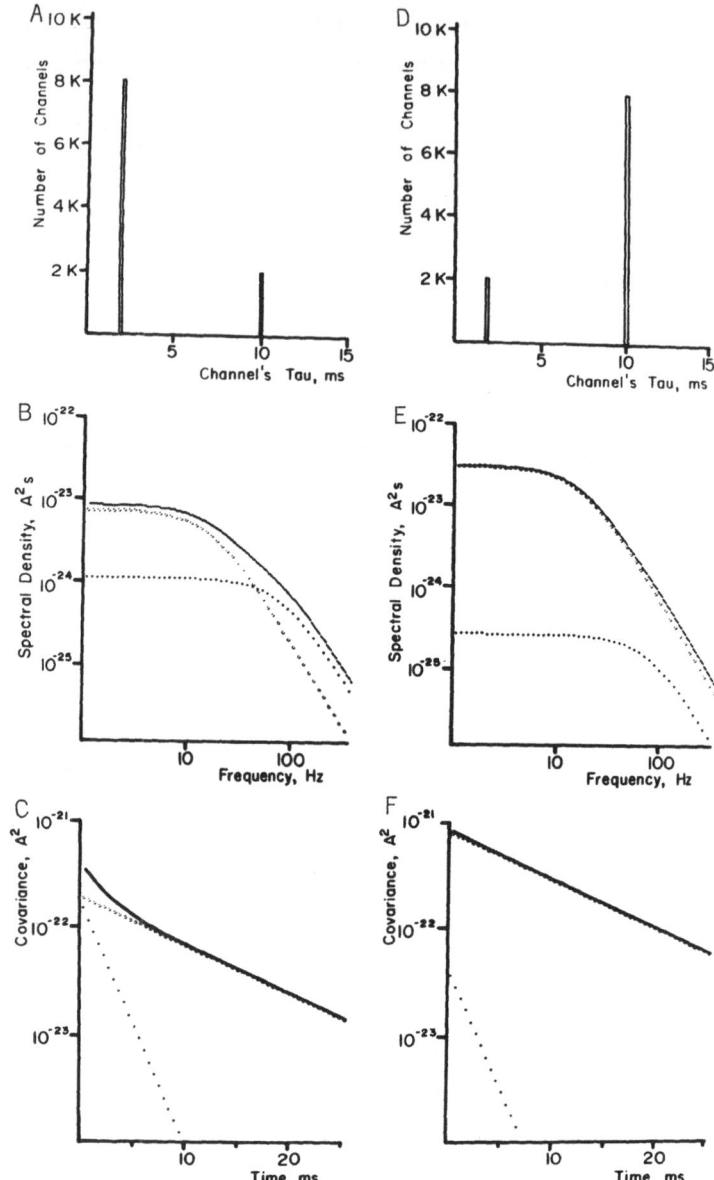

Fig. 2.   Noise analysis of a [hypothetical] heterogeneous channel
          population.   The total number of channels is taken to be
          10,000.   In the case depicted in (A), 80% of the channels
          have a mean open time of 2 ms, and the remaining 20% have
          a mean open time of 10 ms.   The power spectrum which
          would be generated if such channels opened once per

Detecting Channel Heterogeneity

The single channel technique is more sensitive than
fluctuation analysis for measuring heterogeneous populations of
channels, or heterogeneity in the kinetic behavior of individual
channels. For purposes of illustration, I present several
specific situations which, although hypothetical, represent
situations which have been, or might be encountered in biological
preparations. In the first example, two populations of a given
type of channel exist in a cell membrane. The channels are
identical except that they have two different mean open times.
This behavior is seen in the extra-synaptic region of denervated
frog skeletal muscle, for example, where synaptic-type
ACh-activated channels (ACh-channels) exist side by side with
extrasynaptic ACh-channels, which have about a five times longer
mean open time (Sakmann et al., 1980; Hamill and Sakmann, 1981).
I will illustrate here what one would actually measure with

second, had a conductance of 30 pS and a driving force of
100 mV, is shown in (B). The open circles are the
spectrum from just the slow channels, the crosses the
spectrum from the fast channels alone. The solid line is
the sum of the two components. It is what would be
measured in a noise analysis of such a preparation. (C)
is the covariance function which would result from the
same preparation. The symbols have the same meaning as
in (B). (D), (E), and (F) represent the opposite case to
that of (A), (B), and (C): 20% of the channels are fast
and 80% are slow, with mean open times of 2 and 10 ms
respectively. (E) is the power spectrum and (F) is the
covariance function for this case. The symbols have the
same meaning as in (B) and (C).

fluctuation analysis, and with single channel analysis, given this
type of preparation:

The fast and slow channels have mean open times of 2 ms and
10 ms, respectively. Fig. 2A and 2D illustrate two cases in which
the total number of channels is fixed, but the percentage of the
two channel types is different. In one instance there are 80%
fast channels and 20% slow, and in the other case the proportions
are reversed. The power spectra (Fig. 2B and E) and
autocorrelation functions (Fig. 2C and F) for these two cases can
be calculated using several assumptions. Here I have assumed a
total of 10^4 channels, each contributing equally to the signal.
Each channel has a conductance of 30 pS, and opens on the average
once per second, thereby fulfilling the requirement of low
probability of opening. The driving force for currents through
the channel is 100 mV. Fig. 2B shows the power spectrum which
would be generated by 8000 channels with a mean open time of 2 ms
(circles), and that generated by the remaining 2000 channels with
a mean open time of 10 ms (crosses). These spectra are calculated
using the formula:

$$S(f) = \frac{4\gamma(E_m - E_o)<I>\tau_o}{[1 + (2\pi f \tau_o)^2]^{\frac{1}{2}}}$$

where $S(f)$ is the frequency dependent spectral density, γ, E_m, and
E_o have the same meanings as above, $<I>$ is the mean current, and
τ_o is the channel's characteristic open time (Conti and Wanke,
1975; Colquhoun and Hawkes, 1977; Stevens, 1977). The points for
each channel type describe a single Lorentzian curve, which is the
typical form of the spectral density for a channel with a single
open state that has an exponentially distributed lifetime. Since
the current signals through the two channel types are independent
and additive, the fluctuation signals, and thereby the fluctuation

power at each frequency are also additive. The solid line in Fig. 2B is the sum of the two individual components, which would be measured by a fluctuation analysis of this preparation. It clearly consists of two individual components. These data would strongly suggest two kinetic components in the absence of any other information about its preparation.

However, the method of fluctuation analysis is not always sensitive to the presence of two separate kinetic components. If the proportions of channels in Fig. 2A were reversed (as in 2D), such that 20% of the channels were of the fast type and 80% of the slow type, the presence of two components would not be detectable because the signals from the fast channels are swamped by those from the slow. Fig. 2E illustrates the power spectra that would be measured in this case. The fluctuation power of the fast type channel is two orders of magnitude less than that of the slower channel (note that the spectra are displayed on a log-log scale). Because of this large disparity between the two components, the sum of the two components (the solid line in 2B) is nearly identical to the single component spectrum of the slower channel, even at frequencies well above the corner frequency of either individual spectrum. Furthermore, the frequency region with the greatest difference between the single slow-channel Lorentzian and that for both channels is the range where data are least accurate, due to instrumentation noise and statistical errors in the fluctuation measurement. In an actual experiment, a second population of channels comprising 20% of the total would be extremely difficult to detect.

Two components are detectable in one case but not in the other because power spectra are scaled by the square of the channels' mean open time. The normalizing term of the power spectrum includes factors for the mean current and the mean open time, both of which scale the overall signal power. The mean

current, however, is directly proportional to the time the channel
spends in the open state at the low concentration limit (when
channel open time is short compared to the time between openings).
This is represented as:

$$<I> \cong N\gamma(E_m - E_o)\tau_o/\tau_c$$

where $<I>$, N, E_m, E_o, γ, τ_o are as previously defined and τ_c is
the channel closed time. The scaling factor of the power spectrum
is therefore proportional to both the square of the mean open
time, and the number of channels which have that mean open time.
Individual channel openings which are longer carry more current
per event, and therefore, they have more power. When these
channels also predominate in numbers, their contribution to the
overall fluctuation signal is disproportionately large, since
fluctuation analysis measures the power of the signal.

The scaling of the contributions of channels with different
mean open times is a phenomenon which is common to all types of
fluctuation analysis. Figure 2C illustrates the analogous effect
of scaling on the signals measured with auto-correlation analysis
(Conti and Wanke, 1975; Colquhoun and Hawkes, 1977). Fig. 2C is
the expected auto-correlation amplitudes of the fluctuations
caused by 8000 channels with a mean open time of 2 ms and 2000
channels with an open time of 10 ms (displayed on a semi-log scale
as a function of time). The open symbols are the expected values
for the fast channels. The slope of a straight line drawn through
these points is equal to the mean channel open time. The crosses
are the amplitudes for the slow channels, and the solid line, as
above, is the sum of the two components. In this case, the two
different slopes of the auto-correlation curve can again be
detected easily. In Fig. 2F the proportion of channels has been
reversed, and it is no longer possible, at least within the
constraints of actual data, to detect the second component.

It is instructive to examine the results of a hypothetical single channel measurement in this situation, and compare it to the results of fluctuation analysis as discussed above. A different underlying assumption is made when individual events are measured in this type of preparation using the single channel technique. Histograms are made from many observations, and the events are counted as single units with no weighting that is dependent on their length. Since the data are not scaled by the power carried by individual events, the amplitude of two different kinetic components is directly proportional to the fraction of channels which each component represents. This is illustrated in Fig. 3, which shows the log of a histogram plotting the number of occurrences of each duration against time, where the total number of hypothetical observations is 10,000. In Fig. 3A, as above, 80% of the channels, and therefore 8000 events were from channels with a mean time constant of 2 ms (circles), while 2000 events were from channels with a 10 ms time constant (crosses). The solid line is the sum of the two components and represents the data one might measure from such a preparation. Two slopes can be clearly seen, and it could be concluded directly that two kinetic components were present. Fig. 3B shows the expected result of a similar analysis when the proportions of the two channel types are reversed. In this case the two slopes can still be clearly distinguished, in contrast to the results of fluctuation analysis illustrated in Fig. 2.

As a further example of the difference between single channel recording and fluctuation analysis, one can hypothesize the general case in which each individual channel in a preparation has its own characteristic mean open time, τ_o. However, unlike the normal case where all channels are identical, the individual τ_o values are distributed over a given range for a population of channels. The hypothetical situation of a normally distributed population of channels is useful because it further illustrates

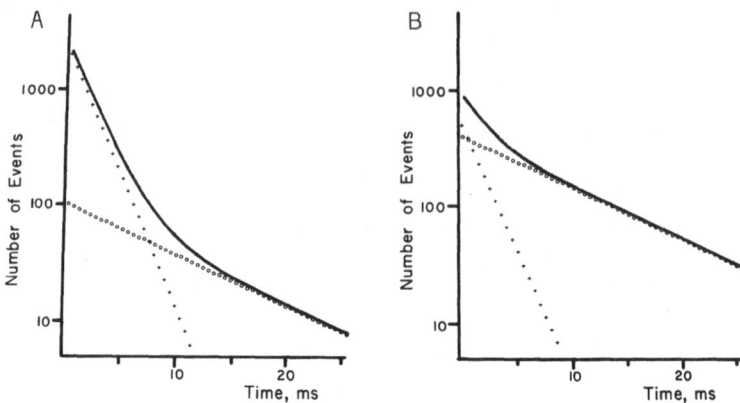

Fig. 3.   Single channel analysis which would result from the
          heterogeneous channel populations in Fig. 2. The total
          number of observed events is taken to be 10,000. The log
          of the number of instances in which a single channel
          opening of each duration is plotted. (A) represents the
          expected observations in the case represented by Fig. 2A:
          the open circles are the events which would be due to the
          slow population of channels, and the crosses are those
          which would be generated by the fast population. The
          solid line is the sum of the two components, and
          represents the actual histogram which would be measured
          in this case. (B) is the corollary analysis for the case
          represented by Fig. 2D. The symbols have the same
          meanings as in (A).

the limitations of fluctuation analysis and the power of single
channel recording.

    Fig. 4 shows the power spectrum which would be produced by a
population of 10,000 channels each of which has its own
characteristic τ_o. The relative proportions of channels with a
given τ_o are distributed around a mean value according to a

Gaussian distribution. This proportioning is illustrated in Fig. 4A, where the number of channels with each τ_o is plotted as a function of that τ_o. The mean τ_o is 20 ms, and the standard deviation about this mean is 5 ms. The total number of channels is 10,000. Using the same assumptions as in Fig. 2, i.e., that channels open once per second, have a conductance of 30 pS, and a driving force of 100 mV, a power spectrum can be calculated by adding all the spectra from individual channels. The resultant spectrum is illustrated as the open circles in Fig. 4B. The solid line in the figure is the spectrum that would be expected from the fluctuations produced by 10,000 channels, each of which had a characteristic τ_o of 20 ms. Although the curves are not identical, they are so similar that it would be impossible to distinguish them in practice. Therefore, one might make fluctuation measurements on a preparation which has a fairly wide distribution of channel kinetics and not be able to distinguish it from a single component system.

Single channel measurements, however, would be able to detect such a distribution. The histogram of channel open times during recordings when it could be assured that only one channel were functioning (such as during bursts, as described below), can be fitted and a τ_o for that channel's open time can be determined. If this analysis is repeated for many channels, then a homogeneous population of channels would, ideally, always yield the same τ_o. However, such an analysis for the type of preparation illustrated in Fig. 4A would yield a distribution of measured τ_o. One could then construct a histogram consisting of the number of instances where a particular value of τ_o was measured. Such a hypothetical histogram is illustrated in Fig. 4C. The axis and form of the histogram are similar to the curve in Fig. 4A which characterizes the population.

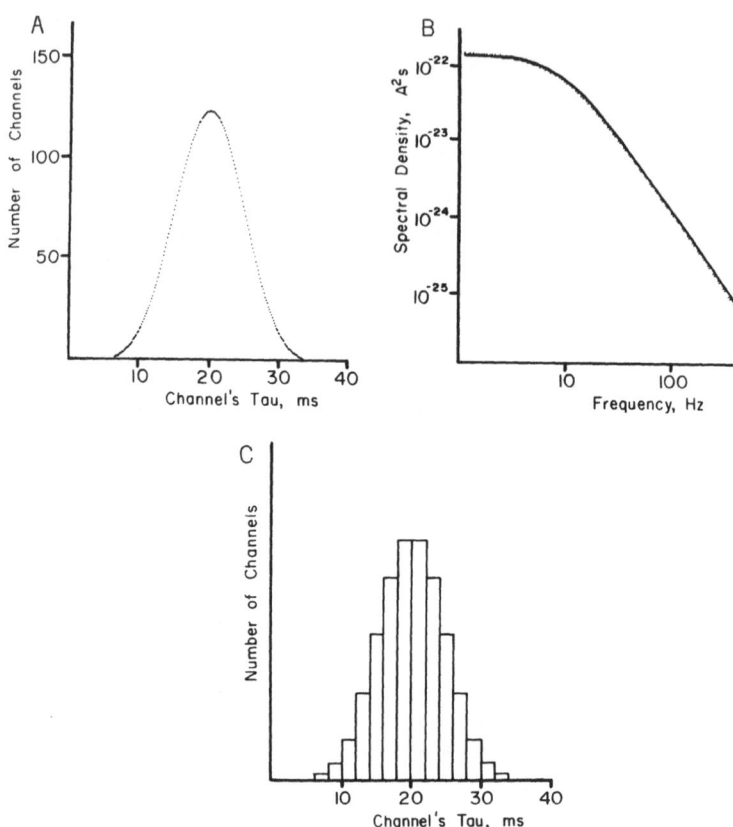

Fig. 4. Noise and single channel analysis for a population of
channels with a distribution of mean open times. (A) is
a histogram of a hypothetical population of 10,000
channels where each channel has a characteristic mean
open time, τ_o, but where the exact value of τ_o is
distributed around a mean value for the whole population.
The mean value of τ_o was 20 ms, and the standard
deviation 5 ms. (B) is the power spectrum which would be
generated by such a population of channels (open circles)
and the power spectrum which would be generated if all
channels had a characteristic τ_o of 20 ms. (C) is a
representation of the result of many different

Although I have presented only cases where the heterogeneity is between different channels, the same principles would apply to the observation of heterogeneity in the kinetics of any one channel. For instance, if a channel had two open states, one short- and the other long-lived, the single channel technique would be more sensitive than fluctuation analysis for resolving the fast component. Furthermore, it would be possible to distinguish between the two types of heterogeneity that I have discussed by using the single channel technique to record under conditions where only one channel was active.

The above discussion illustrates that the single channel technique is not only more sensitive, but also that it provides more information than fluctuation analysis. The cases presented here show that, if one were attempting to deduce the underlying mechanism of channel behavior using fluctuation analysis, one could be deceived into concluding that these systems were simpler than they actually were.

Channel Closed Times

The analysis of channel closed times is much more complex than that of the open times because most channels appear to have more than one closed state. Since the measurement of the closed state is actually the absence of information, there are no direct indications of which state the channel is in at any particular time. If a channel has more than one closed state, then the

measurements of channel open time, where the channel's τ_o is determined under conditions where only one channel is active (either by burst analysis as in Fig. 1, or by analysis of patches with low channel density.)

distribution of all observed closed times should be multiexponential. Furthermore, if more than one channel exists in a patch, then the scaling of the time constants is proportional to the number of channels present. This number must be determined before the exact time constants can be determined.

Despite these complications, single channel recording can be used profitably to give information about the transition rates between closed states, or between the closed and open states. Such information can be obtained by determining the number of channels present (as discussed below), or adjusting conditions so that only one channel is active. If individual components of a multi-component exponential have sufficiently well separated time constants (usually by a factor of three or more), the individual components can be unambiguously fitted. For many channels these conditions can be satisfied, and thus useful information can be obtained.

An example of a channel with a simple open time, but a complex closed time is the ACh channel of the neuromuscular junction. As many as five different components to the closed time can be observed with single channel recording. These are best explained with reference to a simplified state diagram of the channel:

$$2 \text{ ACh} + \text{R} \underset{\longleftarrow}{\overset{\longrightarrow}{}} \text{ACh}_2\text{R} \underset{\longleftarrow}{\overset{\longrightarrow}{}} \text{ACh}_2\text{R*} \underset{\longleftarrow}{\overset{\longrightarrow}{}} \text{D}_1 \underset{\longleftarrow}{\overset{\longrightarrow}{}} \text{D}_2 \underset{\longleftarrow}{\overset{\longrightarrow}{}} \text{D}_3$$

where ACh is the agonist molecule, acetylcholine, R is the receptor, and * indicates the open conformation. The three states, D_1, D_2, and D_3 are desensitized states of the channel. Although the state diagram for desensitization is more complex than the one illustrated here, I have chosen to illustrate a linear scheme for ease of presentation.

The five different components of the closed time are as
follows: 1) <u>Channel reopening before unbinding</u>.   Many of the
individual channel openings have very short interruptions with
mean duration in the range of 200 µs.   They are apparently
instances when the open channel returns to the ACh_2R state, and
then reopens.   Such transitions have been documented by Colquhoun
and Sakmann (1981) and Dionne and Leibowitz (1982) and others, and
are referred to as <u>Nachschlag</u>, or literally, second helping.

2) <u>The dissociated, 'normal' closed state</u> of the channel has
a mean lifetime that is dependent on the agonist concentration
(Sakmann et al., 1980).   At concentrations of ACh of about 10 µM,
this closed time, about 18 ms, is only slightly longer than the
open time of the channel.   It is almost 100 times slower than the
<u>Nachschlag</u> closing, and therefore easily separated from this fast
component.

3) <u>The first 'desensitized' state</u> is a component of closed
times that has a mean of about 1 s, about 50 times slower than the
normal closed state (Sakmann et al., 1980).   This state is often
observed at ACh concentrations that cause moderate desensitization
(about 10 µM).   This closed time can be clearly seen as the time
between 'bursts' of activity, as illustrated in Fig. 5B.   Its
voltage and agonist concentration dependence have not been
carefully studied.   This slow component could be explained by the
presence of a desensitized state, D_1, that undergoes reversible
transitions to and from the active states.

4) A much <u>longer lived desensitized state</u>, with a mean closed
time on the order of 30 s also appears quite regularly at agonist
concentrations of 10-20 µM (Sakmann et al., 1980).   This would
imply that another desensitized state, D_2, exists.

5) Finally, <u>very</u> <u>long</u> <u>closed</u> <u>times</u> occur in recordings of ACh channels after prolonged exposure to desensitizing concentrations of agonist (unpublished observations by the author). Such long closed times are very difficult to study because they are often as long at the patience of the investigator, but if this is indeed a separate component, then still another desensitized state, D_3 is implicated.

The extreme separation of the closed time components in ACh channels has another beneficial consequence. The channel opening events are associated with one another into groups. These associations allow interpretation of whole sets of opening events in terms of state models. In the example given above, the channel open time and the 'normal' closed time are approximately comparable at 10 μM ACh, as shown in Fig. 1. The channel's responses are often limited to bursts of activity which are delineated by pauses of the third category above as shown in Fig. 5B. The length of such bursts can therefore be interpreted as the time that the channel is not in the desensitized states. The distribution of burst lengths would give accurate information about the rate constant for entry into state D_1. Fig. 1A is another illustration of such a burst.

Other types of patterning also exist in ACh channel responses. The association of several openings separated by nachschlag closings, and the association of several bursts into 'clusters' are other examples. The reader is referred to the original references for more information on these phenomena. Numerous other examples of such patterning also exist in the literature. Ca-activated K+ channels (Marty, 1981), and Ca channels (Reuter et al., 1982) are but two examples of channels whose activity appears in bursts.

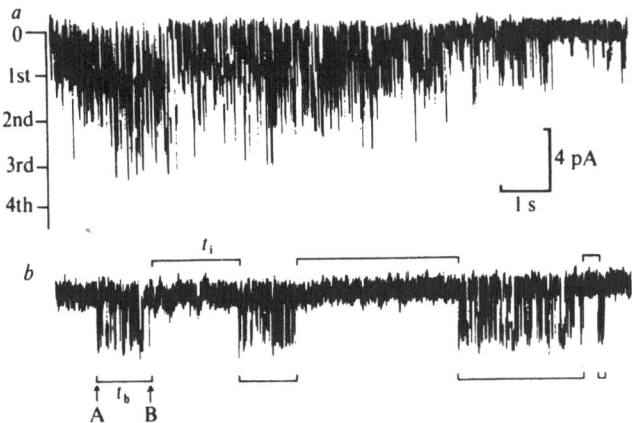

Fig. 5. Multiple components of channel closed time. Application of a patch electrode with 5 μM ACh to extrasynaptic regions of a denervated frog cutaneous pectoris muscle caused a high level of channel activation, which slowly decayed (desensitization). The initial activation, Fig. 5A, demonstrated that 3–4 channels were present in this patch. After several seconds, activity had decreased to sporadic bursts of channel openings and closings, separated by much longer closed intervals as shown in Fig. 5B. Both the length of the closed intervals, t_i, and of the bursting periods, t_b, provide information about the channel's kinetics. Other recording conditions were the same as in Fig. 1. (From Sakmann et al., 1980).

Overall Measures of Channel Kinetics

The channels that have been discussed up to this point have shared the property of stationarity (the probability of opening does not change with time), which has simplified their kinetic analysis. Many channels of interest are non-stationary (their

probability of opening changes with time). However, single
channel recording may still be useful to study such kinetics.
This is illustrated be recordings of Na channels in cultured rat
muscle (Horn et al., 1981). After establishment of a gigohm seal,
the membrane patch was excised from the cell, and the cytoplasmic
side of the membrane was bathed with isotonic CsF. The membrane
voltage was held at -110 mV, and then stepped to -40 mV in order
to activate the Na$^+$ channels. No channel activity was observed at
-110 mV. After the start of the depolarizing pulse, channels
opened stochastically and stayed open for variable periods, as is
shown in Fig. 6A. Occasionally two or more channels opened
simultaneously, indicating that more than one channel was present
in the patch. The pattern of channel opening was different for
every pulse, but during a pulse series, trends in the channel
behavior were apparent. While the single channel conductance,
about 15 pS, stayed the same for channels at all times, the chance
that a channel would open was higher at the start of the pulse
than at the end.

In order to summarize the behavior of the channels within the
patch as a function of time, or to compare this behavior to the
signals produced by many channels functioning simultaneously, the
currents from a set of identical and sequential pulses ('pulse
set') were averaged, thereby determining the mean current as a
function of time. Since it is assumed that channels in a membrane
function independently, and that their currents add to give the
total signal, the mean currents from many responses of one channel
are analogous to the currents which would be obtained if several
hundred channels were stimulated at once, as in whole cell
recording (Assuming that all other conditions, including the
stimulation frequency, were the same).

The calculated mean currents are directly proportional to the
probability that any individual channel will be open as a function

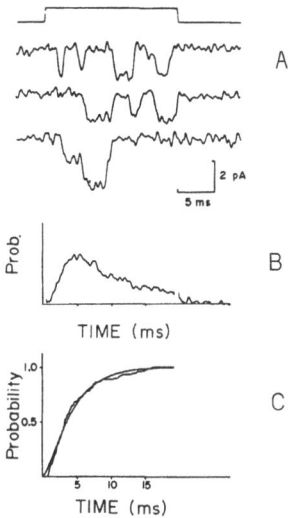

Fig. 6.   Single channel recordings in an excised patch of tissue
           cultured rat myotubes at 8.9°C. (A) shows three
           individual records from a set of 144 depolarizations to
           -40 mV from a holding potential of -90 mV. The voltage
           pulse is depicted above the current traces. In the third
           trace the current from two channels overlapped. (B)
           shows an arbitrarily scaled and inverted sum of the
           currents from the same set of depolarizations. It
           estimates the probability density that a channel is open
           at a given time. (C) plots the cumulative distribution
           of times after the start of the voltage step to the first
           opening for the same set of voltage pulses. The smooth
           curve is a theoretical histogram for a sequential model
           of activation which has two closed states leading to an
           open state. (From Horn et al., 1981).

of time, since channels are either open or closed and always carry
the same current. One proportionality factor is the number of
channels in the patch. Fig. 6B illustrates the mean current from

144 pulses such as those in Fig. 6A, plotted as the probability of a channel being open as a function of time. The probability-of-being-open curve has many of the characteristics of a Na^+ current from a voltage clamped cell. It rises quickly after a slight delay, reaches a peak, and then returns to the zero level, inactivating. As in macroscopic recordings, the shape of this curve carries kinetic information about the function of the channel. For example, the shape of the rising phase can be compared with the predictions of specific models to determine the number of individual steps which precede the opening of the channel, while the time course of the falling phase gives information about the rate of inactivation. The probability scale has been left unlabeled, however, because the exact number of channels in this patch of membrane was unknown. In order to use the information obtained by determining the average of many channel responses to pulses, the exact scaling for the resulting probability curve must be determined. To do this one must estimate the number of channels within the patch, as will be discussed below.

Fig. 6C illustrates another kind of measurement which can be made on these data. After the start of the pulse, a variable time occurs before a channel opens. This period is referred to as the latency to first opening, and it gives an indication of the duration of the kinetic processes which precede the channel's opening. When this quantity is measured for each pulse a histogram can be constructed. Alternatively, the cumulative histogram can be calculated, as shown in Fig. 6C. This measurement plots the probability that an opening will already have occurred as a function of time. The shape of this curve gives kinetic information about the channel which differs from that obtained from the mean current traces. It does not include kinetic contributions from the length of time that the channel remains open, or from subsequent inactivation processes, because the measurement ends with the first channel opening. However, the scaling of the cumulative latency curves is also influenced by the number of channels which are present in the patch, since it measures the time it takes for any one of the channels in the patch to open. Therefore it is important, as with mean current measurements, to estimate the number of channels in the patch.

Determining the Number of Channels in a Patch

Sometimes one can estimate directly the number of channels under the patch. For instance, if channels can be activated with a very high probability of opening, then the maximum number open can be determined for a data set. The greatest number of channels simultaneously open will probably be the number of channels under the patch. However, this procedure is not always feasible.

In our studies of the Na channel, the maximum probability of opening for any one channel was relatively low at -40 mV, about 0.15 (Patlak and Horn, 1982). For greater depolarizations this probability increased, but so did the rate with which the channels opened and closed, and this limited resolution at any finite recording bandwidth. Furthermore, as the size of the depolarization increased, the single channel currents decreased, while the size and length of the switching transients increased. This further reduced the possibility of resolving single channel transitions, especially near the start of the pulse when the probability of opening was greatest. We therefore developed a method with which we could determine accurately the number of channels in a given patch under these conditions.

A reasonable estimation of the number of channels can be made using a binomial analysis of the data set. Neher and Sakmann (1976) used such an analysis to show that currents which reached the second quantum level were produced because two distinct

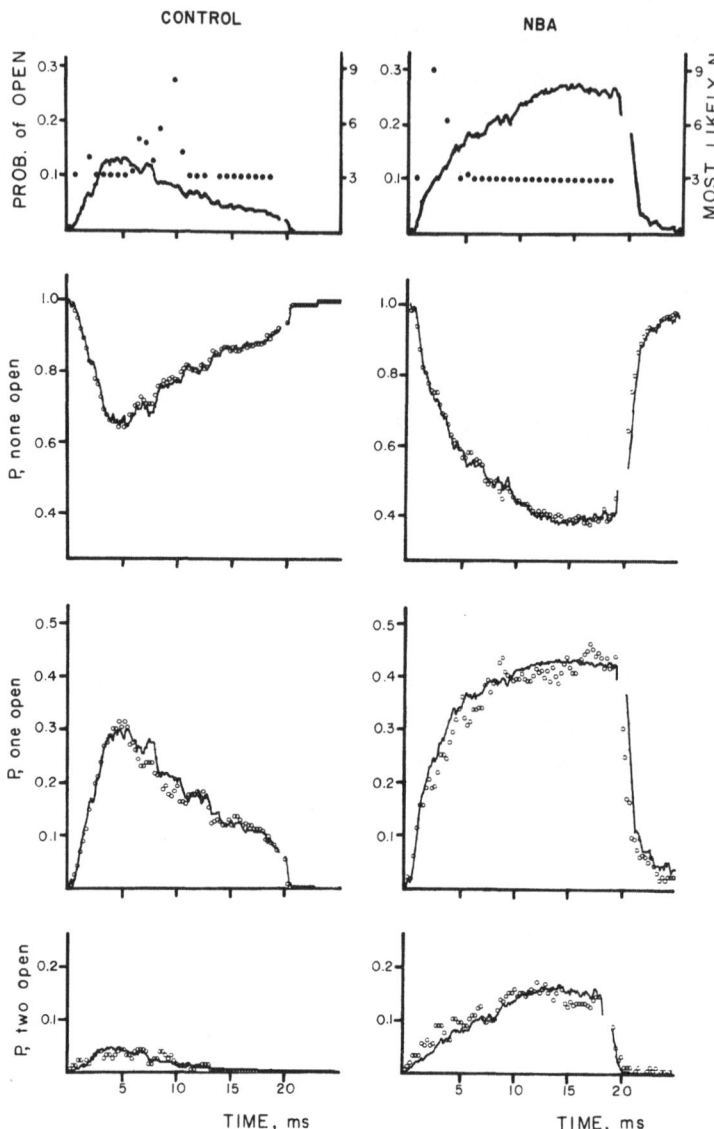

Fig. 7.   Binomial analysis of Na channels from a patch of rat
          myotube in tissue culture before and after treatment with
          N-bromoacetamide, which removes inactivation.   The open
          symbols in the lower three sets of graphs plot the
          observed probabilities that zero, one or two channels
          were open following the start of the pulse.   A total of

channels were open simultaneously. Basically, the binomial
hypothesis says the following: If many independent events each
have the same probability of occurring, the probability that x
will occur simultaneously is given by the formula:

$$P_x = [N!/(x!(N - x)!)] \, p^x \, (1 - p)^{N - x}$$

where P_x is the probability of x simultaneous occurrences, N is
the number of independent entities, and p is the probability of
any one event. Applied to single channel data, the chance of
observing x simultaneously open channels when N channels are
present is given by the above equation. Since the number of
channels open at any particular time after the start of the pulse
can be observed directly, curves can be constructed of these
probability levels as functions of time.

Such an analysis is shown in Fig. 7. The open circles in the
lower panels are the measured probabilities that zero, one, or two

144 pulses were analyzed. The pulse potential was −40 mV
and the holding potential was −110 mV. The temperature
was 10°C. The solid circles in the top panels are the
values for N, the number of channels under the patch,
which were most likely to have given the observed
probability distributions. The best estimates for these
patches were 3 channels. The solid lines in the top
traces are the mean currents from these pulse sets. The
solid lines in the lower graphs are the probabilities of
zero, one, or two channels being open, given the observed
mean currents and binomial distribution with three
channels. (From Patlak and Horn, 1982).

channels were open simultaneously at each time after the start of
the pulse. The mean current, as measured above, is also a
function of the number of channels, N, and the probability, p. In
principle, data points from any two curves at a particular time
can be used to solve the two equations for the unknowns N and p.
In practice, the range of N and p which we observed for Na^+
channels was such that even with sets of several hundred pulses
the scatter in the calculated points was very large, making the
calculations unreliable. In order to determine the best estimate
for the number of channels it was necessary to use all the data,
and to determine quantitatively the likelihood that the observed
distribution could have been generated by a given number of
channels. The likelihood function is given by Dahiya (1981):

$$L(N) = \prod_{i=1}^{m} \frac{N!}{X_i!(N - X_i)!} \left(\frac{<I>}{N}\right)^{X_i} \left(\frac{1 - <I>}{N}\right)^{N-X_i}$$

where L(N) is evaluated over m observations, and X_i is the level
of the i_{th} observation at any particular time. This determination
of likelihood was repeated for all reasonable values of N. The
value of N which yields the maximum likelihood is the best
estimate for N at that particular time. The solid points in the
top panels are the numbers of channels giving the maximum
likelihood for each time point. The solid lines in the top panels
are the calculated probabilities of a channel being open based on
a best estimate of 3 channels for these j patches. The solid
lines in the other panels are the probabilities of zero, one or
two events based on the overall probability of the top panels and
the binomial equations. The good overall fit of the solid lines
to the open circles indicates that the maximum likelihood estimate
is reasonable for all times. Determination of the number of
channels in a particular patch makes the normalization of mean
currents and latency curves plausible, and allows quantitative

measurement of a channel's rates.

STATISTICAL ANALYSES OF SINGLE CHANNELS

Using single channel analysis, it is possible to determine the relative frequencies with which particular events happen, and to draw conclusions about channel kinetics from these measurements. I will discuss three examples of such analyses: 1) Use of binomial analyses to determine appropriate state diagrams for channels, 2) Maximum likelihood analysis to determine rate constants from complex single channel data, and 3) Conditional probability analysis to test possible inactivation models.

Binomial Analysis of State Diagrams

Labarca and Miller (1981) have used single channel recording to show that K^+ channels from frog sarcoplasmic reticulum have three sequentially arranged states, two of which are conducting. They observed two distinct levels of conductance, 50 and 150 pS, when channels from sarcoplasmic reticulum were inserted into artificial bilayer membranes. Two alternate schemes could give rise to this observation. Either there were two independent channels with conductance of 50 and 100 pS, or one channel had two conductance states. They were able to rule out the former model because they never observed transitions from zero to 100 pS. The hypothesis that these were independent channels predicts that such transitions should occur fairly often, so this explanation could be ruled out.

These investigators also wanted to determine the most appropriate state diagram for the three possible states, closed

(C), open with 50 pS conductance (α), or open with 150 pS conductance (β). Two possible configurations were proposed to explain these states:

(I) $C \underset{\longleftarrow}{\longrightarrow} \alpha \underset{\longleftarrow}{\longrightarrow} \beta$ (II) $C \underset{\longleftarrow}{\longrightarrow} \alpha$

Labarca and Miller measured the probability that a given transition would occur (i.e. between C and α, C and β etc.) and compared these values with those predicted by models I and II above. Transitions between state C and state β were found to be very rare. The few instances of such transitions were equal to the probability that the state α was occupied for a short time during the transition, but was not occupied long enough to be observed. These data were consistent with model (I) and led the authors to conclude that this was the most likely state diagram to explain the channel's kinetics.

Both aspects of this analysis, the determination that one channel had two conductance states, and that the states were arranged in a sequential fashion would have been much more difficult to obtain using macroscopic or fluctuation analysis. However, these are straightforward measurements when measuring single channel signals.

Maximum Likelihood Analysis of Channel Kinetics

Horn and Lange (1983) have extended the analysis of non-stationary single channel data by developing a maximum likelihood analysis that takes into account the time sequence of the data, not just the probabilities for observing a given current level at one time. They have implemented the technique because data sets from single Na$^+$ channels are often difficult to handle quantitatively. These difficulties arise from the fact that the

Na^+ channel appears to have more than one open state, there are usually several channels in a patch, the data are non-stationary, and the recordings have finite length.

This method of analysis involves determining the probability that a given piece of data was generated by a particular set of states and transition rates. The rate constants are systematically altered, and those rates that give the maximum likelihood are taken as the best fit to the data. In other words, such an analysis is designed to determine ex post facto what was the most likely model that led to the particular data for any one experiment. Horn and Lange have demonstrated that such a method can be used successfully to analyze complex single channel data generated by computer, and they are currently applying the method to the analysis of Na^+ channels.

Although this method is computationally very demanding, it shows the power of statistical methods for determining rate constants from a given, complicated set of single channel data. Optimization of this technique should make it possible to fully determine the rate constants for all possible transitions from a sample of data that can be easily obtained on a single patch.

Conditional Probabilities

Finally, it is possible, using single channel recording to measure a channel's activity based on its past history. This information can be used to determine conditional probabilities. One example of this type of conditional probability is the determination of the single channel open time, as discussed above. Another example is the determination of the probability of a channel opening if that channel did not open at any time during the first part of a pulse.

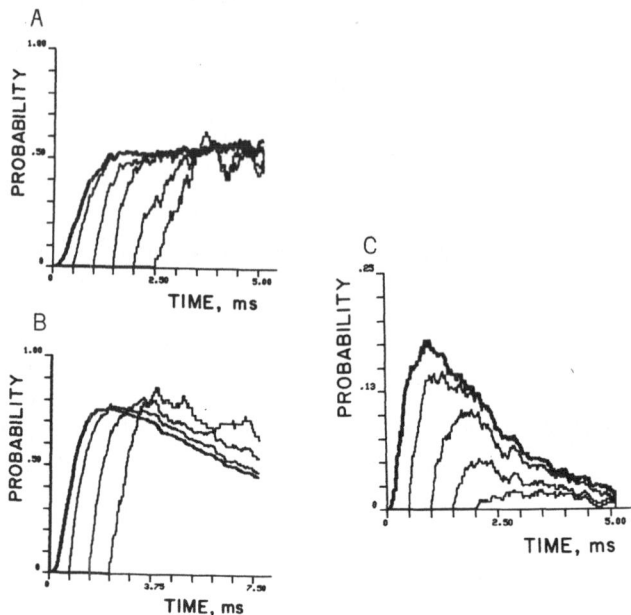

Fig. 8.   Overall and conditional probabilities of being open from
          computer generated single channel signals using three
          models of Na-channel activity. In each case, 1000
          responses of a single channel were mimicked by using
          random numbers to determine jumps between states. The
          dark line in (A) is the probability that a channel
          without inactivation would be open after a pulse to -40
          mV. The channel had 3 closed and one open states. The
          rate constants for transitions were determined using
          Hodgkin and Huxley's equations (1952). The thin lines
          are the conditional probabilities that the channel
          opened, given that it failed to open during the first
          part of the pulse. The five traces are for delay
          intervals of 0.5, 1, 1.5, 2, and 2.5 ms. Fig. 8B is a
          similar analysis from a channel which always passed
          through the open state before inactivating. Rate
          constants were determined from a model presented by
          Armstrong and Gilly (1979). The dark line is the overall

This type of conditional probability was first proposed by Horn et al., (1981) and has recently been discussed in depth by Patlak (1983). It measures the probability that a channel will be open at a particular time after the start of a pulse, given that it did not open during a specified interval at the start of the pulse. If channels can undergo transitions between certain of their closed states during this initial interval, then this will influence subsequent probabilities. If such transitions are not possible, then the conditional probability should be largely independent of the conditioning interval.

Figure 8 shows several examples of this type of measurement for data that was generated using several computer models of Na channel activity (Patlak, 1983). In Fig. 8A, activation of channels took place via a series of three sequential closed states. No inactivation was present. The dark trace shows the probability-of-being-open curve for 1000 pulses with one channel. The subsequent lighter traces are the conditional probability curves for several delay intervals. These traces are very similar to the overall probability. Time dependent transitions between the several closed states during the conditioning period change only the initial part of the curve. Figure 8B shows the conditional probabilities expected if the channel inactivates only after reaching the open state. Again, the conditional probabilities (light traces) are largely a time shifted replica of the overall probability (dark trace), since it does not matter how long the channel remains in the closed states before it activates.

probability. (C) is the analysis of a Hodgkin-Huxley channel with inactivation that was independent of activation. The conditional probability diminishes because channels can inactivate before they open. (From Patlak, 1983).

However, channels that can inactivate from their closed
states have a chance to inactivate before their first opening.
Since channels that stay closed longer will have a greater chance
to inactivate before they first open, the conditional probability
will decrease as its delay interval becomes longer. Fig. 8C shows
the expected conditional probability curves for the case of fully
independent inactivation. In this case the conditional
probabilities are quite different from the overall probability.
The diminution of the former is due to inactivation of channels
before they inactivate. Models in which inactivation is partially
coupled to activation give results intermediate between those
shown in 8B and C, allowing the possibility of fitting a proposed
theory to actual observed data using this method.

Horn et al. (1981) used such analysis for the Na$^+$ channel to
show that channels can inactivate without opening. Conditional
probability measurements, therefore, are yet another way in which
single channel data can be used to provide unique information
about the function of channels.

CONCLUSION

I have outlined a number of different methods of extracting
information about channel kinetics from single channel data. Each
method measures something different, and therefore all contribute
independent pieces of information which help determine a kinetic
model for the channel. Possible measurements include the duration
of the channel's open and closed times, the number of exponential
components in the channel closed times, the mean current, latency
to first event, and the duration of activity that is separated by
particular components of the closed time (burst analysis).
Furthermore, I have discussed techniques for determining the
number of channels in a patch, and for measuring the

appropriateness of a particular state diagram. Although these
different measurements may not be enough to completely resolve all
rates of a channel's activity, measurements of single channel
activity provide a greater degree of specificity in determining
particular rate constants than do measurements of mean current or
fluctuation alone.

I have also discussed other advantages of recording single
channel signals, such as the ease of measuring single channel
conductance, or of detecting heterogeneity in channel populations
or in channel kinetics. Combined with the advantages listed in
the preceding paragraph, the argument for recording from single
channels is overpowering, when the extra amount of information
that this method provides is needed.

However, this extra information is not free. The method is
usually experimentally more demanding than other recording
techniques. Recordings must be long to avoid statistical errors,
and data analysis often is much more complex and time consuming
than with other methods. It is therefore necessary to determine
the easiest way to obtain the information that is needed. For
example, fluctuation analyses are quite valid for many types of
channels. In order to examine the influence of an experimental
maneuver on a channel's open time, it may be considerably more
efficient to collect fluctuation data, rather than the detailed
single channel data. The basic assumptions that underlie
fluctuation analysis could be checked by examining the single
channels, but all other measurements need not be at this level.

Finally, single channel recording will never replace the
methods of whole cell recording, voltage clamping, or fluctuation
analysis. Indeed, the patch clamp and bilayer membrane techniques
make extensive use of all of the above methods, and will continue
to do so, because they provide valuable information. Recording

the individual openings from single channels simply gives an
additional source of information that can help bring us to the
goal of understanding channel activity.

REFERENCES

Anderson, C. R., and Stevens, C. F., 1973, Voltage clamp analysis
    of acetylcholine produced end-plate current fluctuations
    at frog neuromuscular junction, J. Physiol., 235:655.
Armstrong, C. M., and Gilly, W. F., 1979, Fast and slow steps
    in the activation of Na channels, J. Gen. Physiol., 74:691.
Bean, R. C., Shepherd, W. C., Chan, H., and Eichner, J. T., 1969,
    Discrete conductance fluctuations in lipid bilayer protein
    membranes, J. Gen. Physiol., 53:741.
Colquhoun, D., and Hawkes, A. G., 1977, Relaxation and
    fluctuations of membrane currents that flow through drug
    operated channels, Proc. R. Soc. Lond. B., 199:231.
Colquhoun, D., and Hawkes, A. G., 1981, On the stochastic
    properties of single ion channels, Proc. R. Soc. Lond. B.,
    211:205.
Colquhoun, D., and Sakmann, B., 1981, Fluctuations in the
    microsecond time range of the current through single
    acetylcholine receptor ion channels, Nature, 294:464.
Conti, F., and Wanke, E., 1975, Channel noise in nerve membranes
    and lipid bilayers, Quarterly Rev. Biophys., 8:451.
Dahiya, R. C., 1981, An improved method of estimating an integer-
    parameter by maximum likelihood, Am. Statistician, 35:34.
Dionne, V. E., and Leibowitz, M. D., 1982, Acetylcholine receptor
    kinetics: A description from single channel currents at
    at snake neuromuscular junctions, Biophys. J., 39:253.
Gordon, L. G. M., and Haydon, D. A., 1972, The unit conductance
    channel of alamethicin, Biochim. Biophys. Acta, 255:1014.

Hamill, O. P., Marty, A., Neher, E., Sakmann, B., and Sigworth,
    F. J., 1981, Improved patch-clamp techniques for high-
    resolution current recording from cells and cell free
    membrane patches, Pflügers Archiv., 391:85.

Hamill, O. P., and Sakmann, B., 1981, Multiple conductance states
    of single acetylcholine receptor channels in embryonic
    muscle cells, Nature, 294:462.

Hladkey, S. B., and Haydon, D. A., 1972, Ion transfer across lipid
    membranes in the presence of Gramicidin A. I. Studies of the
    unit conductance channel, Biochim. Biophys. Acta, 274:294.

Hodgkin, A. L., and Huxley, A. F., 1952, A quantitative
    description of the membrane current and its application to
    conduction and excitation in nerve, J. Physiol., 117:500.

Horn, R., and Lange, K., 1983, Estimating kinetic constants from
    single channel data, Biophys. J., 43:207.

Horn, R., Patlak, J., and Stevens, C., 1981, Sodium channels need
    not open before they inactivate, Nature, 291:426.

Katz, B., and Miledi, R., 1970, Membrane noise produced by
    acetylcholine, Nature, 226:962.

Labarca, P. P., and Miller, C., 1981, A K^+-selective, three
    state channel from fragmented sarcoplasmic reticulum of frog
    leg muscle, J. Memb. Biol., 61:31.

Marty, A., 1981, Ca-dependent K channels with large unitary con-
    ductance in cromaffin cell membranes, Nature, 291:497.

Neher, E., and Sakmann, B., 1976, Single channel currents
    recorded from membrane of denervated frog muscle fibres,
    Nature, 260:799.

Neher, E., Sakmann, B., and Steinbach, J. H., 1978, The
    extracellular patch clamp: A method for resolving currents
    through individual open channels in biological membranes,
    Pflügers Archiv, 375:219.

Patlak, J. B., 1983, Conditional probability measurements on two
    models of Na channel kinetics, in: "The Physiology of
    Excitable Cells," A. Grinnell and W. Moody, editors. Liss,
    Inc., New York.

Patlak, J. and Horn, R., 1982, The effect of N-Bromoacetamide on
    single sodium channel currents in excised membrane patches,
    J. Gen. Physiol., 79:333.

Reuter, H., Stevens, C. F., Tsien, R. W., and Yellen, G., 1982,
    Properties of single calcium channels in cardiac cell
    culture, Nature, 297:501.

Sakmann, B., Patlak, J., and Neher, E., 1980, Single acetyl-
    choline-activated channels show burst-kinetics in presence
    of desensitizing concentrations of agonist, Nature, 286:71.

Stevens, C. F., 1977, Study of membrane permeability changes by
    fluctuation analysis, Nature, 270:391.

MEMBRANES AND CHANNELS

PHYSIOLOGY AND MOLECULAR BIOLOGY

Robert S. Eisenberg

Department of Physiology
Rush Medical College
Chicago, IL   60612, USA

INTRODUCTION

Most of the papers in this book discuss the properties and
roles of channels in membranes, and the methods needed to
investigate them. Work on channels has evolved (in large measure)
from older work on the properties of membranes themselves. Since
channels are the major pathways for solute movement, the mechanism
of solute movement can best be investigated when channels are
embedded in as simple a membrane as possible, attached to as
simple an apparatus as possible. Indeed, that is why single
channel measurements, as described in several chapters in this
book, have created such excitement and are so promising for the
future of membrane biophysics.

Membranes, however, are not just structures in which
interesting channels are found. The structural organization of
membranes and nonuniform distribution of channels within membranes
are also of considerable biological interest since both play an
important role in tissue function. This paper presents a
systematic method for analyzing complex membrane architecture. In

particular, I consider in detail the role of restricted
extracellular spaces, formed by infolded membranes. From the
electrical point of view, these spaces produce a distributed
resistance in series with most of the plasma membrane. From the
electrochemical point of view, these spaces provided a restricted
diffusion space (an 'unstirred layer') in which physiological
current flow produces significant changes in solute concentration.
I argue that the biological consequences of restricted
electrodiffusion are profound and are likely to be widely
exploited in the function of many cells and tissues.

While working on this paper, and thinking about the role of
cell structure in cell function, I attended meetings on protein
structure in honor of Dr. M. Perutz and Prof. J.T. Edsall, the
latter my tutor at Harvard College. Both meetings had many
lectures devoted to structure/function relations of protein
molecules and I was struck by the similarity of these molecular
questions, discussed in the language of chemistry, to questions
concerning structure/function relations of cells and tissues,
usually discussed in the language of physiology. This similarity
seemed hardly a coincidence and led to the discussion presented in
the second section of this paper. I argue there, in a rather
philosophical way, that the investigation of biological systems
has a certain unity, whether those systems occur on a length scale
of nanometers or centimeters. The unity of biological questions
is in contrast to the diversity of the physical sciences, where
the nature of the systems and scientific investigation depends
profoundly on the length scale being studied. The properties of
physical systems are quite distinct, requiring quite different
experimentation and physical description if they are elementary
particles (on a length scale of femtometers), atoms (picometers),
molecules (nanometers), everyday objects (centimeters), or
astronomical objects, ranging from solar systems (10^{12} meters), to

galaxies (10^{21} meters), to the universe (10^{26} meters). The
mathematical laws describing physical phenomena on these different
length scales seem more similar than they really are: when
predictions of practical experimental results are made, the
different distances scales require quite different treatment of
the same 'unified' equations.

I believe biological systems have a great deal of unity
because they are all built by the same evolutionary process, the
the processes of evolution are much the same no matter what the
size of the resulting system. The properties of systems built by
evolution seem to be closely related to the 'chaotic' outputs of
simple mathematical models used to describe idealized evolutionary
systems (May, 1976). I argue that only certain questions can be
profitably and efficiently asked of chaotic systems: "What is
there?" "What does it do?" "How does it work?" "How did it
evolve?" These questions seem to me to be the central ones
confronting biologists, no matter what length scale they study.

PART 1 - RESTRICTED. EXTRACELLULAR SPACES

Role of extracellular spaces. We begin with a discussion of
one aspect of membrane architecture, namely the role of the narrow
extracellular spaces commonly found in nerve and epithelia,
skeletal and cardiac muscle.

The chemical consequences of extracellular spaces occur
because the fluxes of ions across membranes can drastically change
the concentration of ions in the space outside membranes. This
change in solution composition can change both the properties of
channels and the ions which flow through them. Thus, the flow of
ions through channels can be regulated by the geometry of the

extracellular space as well as by the intramolecular properties of
the channels themselves. In this respect, the accumulation of
potassium and depletion of calcium in extracellular spaces will be
most provocative, since the changes in concentration of these ions
is large and the physiological role of these ions is considerable.
In the case of potassium, changes in concentration would modify
the 'resting' potential of neighboring cells, modifying all the
processes which depend on <u>that</u> voltage. In the case of calcium,
changes in concentration would modify the calcium current through
calcium channels; it would modify the conductance of channels to
other ions; and it would act as a signal to the various membrane
enzymes which use extracellular calcium as a trigger signal.

Consider the particular case of cardiac muscle. The natural
cardiac function, the cardiac output of blood flow, is closely
coupled to the duration of the ventricular action potential.
Changes in concentration predicted in narrow extracellular spaces
between ventricular cells are likely to change the duration of the
action potential. Thus, cardiac output is likely to be regulated
by the properties of an extracellular space, as well as by the
properties of membrane channels.

Despite these speculations, relatively little attention has
been paid to the functional role of extracellular spaces, probably
for good reason. They are hard to study, they are not molecular
in nature, and their function is not evident. Indeed, much of the
progress in physiology in this century has occurred in the study
of those tissues without prominent structural complexity. The
classical physiological 'preparations' of axons, red blood cells,
and neurons (at least in their invertebrate peripheral
idealization) may actually have been chosen because of their
relative simplicity. The preparations of interest here

(epithelia, cardiac and skeletal muscle) may have been avoided because of their complexity.

One of the classical physiological preparations bridges the gap between the appalling complexity of syncytial structures, like epithelia, and the appealing simplicity of the squid axon membrane. Skeletal muscle fibers have extensive structural complexities, and yet of a periodic kind. These cells are nearly crystalline in the regularity of their structures. Their membranes form a complex network of tubular invaginations, yet these networks occur (to a good approximation) in a regular repeat. Indeed, the complexity of the network occurs at such a small length scale that it can be averaged away, as a statistical mechanic averages over quantum mechanical states he cannot accurately describe, leaving macroscopic properties relatively easy to deal with.

It was through the study of skeletal muscle that a systematic procedure for analyzing structural complexity was created (by Falk and Fatt, 1964; the role of morphological measurements was clearly stated by Mathias et al., 1977) and the entire procedure was generalized and described in Eisenberg and Mathias (1980).

That procedure has now been applied to a number of tissues and cells, most notably syncytial tissues like heart (e.g., Levis, et al., 1983) and epithelia like the lens of the eye (Mathias et al., 1979), the gall bladder (Clausen et al., 1979; Schifferdecker and Frömter, 1978) or gastric mucosa (Diamond and Machen, 1983). It has proven reasonably successful. The procedures have not yet been applied, at least in any systematic way, to their most obvious target, neurons of the vertebrate central nervous system. Here we must expect a structural analysis to be a prerequisite to

the understanding of neuronal integration, the methods by which
neurons compute their outputs from their many inputs.

Structural analysis of electrical properties. Structural
analysis provides a systematic procedure to analyze tissues and
cells with complex membrane architecture. Structural analysis
starts with a description of the topology of membranes, which
corresponds closely to the mind's eye view of a tissue presented
as line drawings in textbooks. The analysis proceeds to measure
the amounts of membranes and the spaces they enclose, often using
the techniques of stereology. The books of Weibel (1981, 1982)
and the papers of Brenda Eisenberg and co-workers (1983) provide
details of theory and applications in this area. Suffice it to
say here that the problems of measurements of biological structure
can be considered solved, at least the measurement of the volumes
of cells and extracellular spaces and the measurement of surface
areas of membranes. But we must remember that the problem of
defining shape quantitatively -- which is part of the problem of
pattern recognition outside the human nervous system -- is a
formidable task. The problem of measuring that shape is even
harder. Both problems may be greatly simplified if attention is
paid to only that subset of structures which can arise from
self-replicating cells (Dormer, 1980, and Meinhardt, 1979, discuss
some of the consequences of self-replication from a morphological
and mathematical perspective, respectively).

Tissue preservation. An outstanding problem in structural
analysis, indeed in all of morphology, is that of tissue
preservation. No methods are known which preserve both
physiological structure (i.e., the volume of cells and
extracellular spaces) and fine structure as observed in the
electron microscope. Chemical fixation takes far too long, and
involves too many unknown reactions, osmotic effects, and possible

artifacts, to allow confidence in the resulting image of physiological structure.

Rapid freezing better preserves the physiological structure of cells, provided the structures are not more than some 10 μm from the cooling block. But even close to the cooling block, freezing is not rapid enough to guarantee preservation of the volumes of small structures in which volume could change significantly in milliseconds. And, of course, not all fine structure is accessible to this technique.

New techniques of tissue preservation are clearly needed if structural analysis is to fulfill its biological potential. One possibility involves the use of bifunctional photoaffinity reagents to effect light-activated fixation. Consider a molecule which is membrane permeable and biologically inert prior to irradiation. If the compound became a potent protein cross-linker only after absorption of light, one might be able to obtain rapid fixation and good preservation of physiological and fine structure. Admittedly, the chemical and biological requirements which must be satisfied to make this approach useful are many. The compounds must be membrane permeable to very high concentrations, biologically non-disruptive and non-toxic at these concentrations and they must, upon irradiation, cross link with high efficiency and speed. Of course, cross-linking might take milliseconds, being limited not only by the chemistry but also by the time until Brownian motion brings the relevant proteins close enough together to allow cross-linking to occur. But, cross-linking on even that time scale would importantly increase our capabilities. In fact, I suspect that fixation might be quite rapid because I guess that all fixatives fix by cross-linking soluble proteins to nearby cytoskeleton, although this guess should be clearly identified as the unsupported speculation it

really is. If the cytoskeleton is within a few nanometers of all
cell proteins, collisions between it and soluble proteins will
occur very often, and photoactivated fixation might be just as
rapid.

In collaboration with Jeanne Nerbonne (of Cal Tech), we are
presently considering suitable bifunctional photoaffinity reagents
which could be utilized for this purpose. The compounds of
Mikkelsen and Wallach (1976), and other commercially available
reagents seem a logical starting point.

Theoretical predictions. The anatomical description,
topological and morphometric, is not too useful biologically until
its functional consequences can be predicted. That is the role of
theory. A mathematical model of the measured structure is needed
to predict the electrical properties expected from that structure,
with the minimum of assumptions. The prediction, of course, must
involve some assumptions, because of mathematical difficulties and
anatomical complexities, and also because the mathematical model
includes unproven assumptions of membrane properties. Indeed, to
make the analysis tractable, we often must assume that the
membrane properties of interest are linear, i.e., independent of
voltage or other driving forces.

The assumption of linearity. Although excitable membranes
are certainly nonlinear over the extended range in which they
function, an assumption of linearity is not as restrictive as it
seems. The assumption allows analytical solutions of our
mathematical models, analytical solutions which depend quite
simply on membrane architecture, on the extracellular spaces and
membrane capacitances, and thus give qualitative understanding of
the electrical properties of the tissue or cell. In fact, these
analytical solutions can often be cast as equivalent circuits

(with no loss of accuracy) and the power and intuition of circuit analysis thus can be applied to our complex biological tissue, without loss of rigor. This simplification to an equivalent circuit can only be done (in an accurate way, with known mathematical error) when the circuit is linear.

The assumption of linearity is also necessary because it allows a drastic reduction in the number of free parameters in our model. Under the condition of linearity the detailed properties of channels have a much smaller role in determining electrical properties than they do in general. Thus, our ignorance of these detailed channel properties will have a relatively minor effect on our analysis of the properties of the extracellular space. Finally, the assumption of linearity is also more biologically relevant than it may seem to workers on excitable membranes. Most cells have a substantial range of voltage over which their properties do not change. Thus, a linear analysis of electrical properties is also an analysis of the properties of cells in a physiologically relevant state.

Properties of channels do influence linear properties, of course; they determine the resting conductance of membranes and they may have a role in determining a component of membrane capacitance as well (see Bezanilla et al., 1982; Fernandez et al., 1982, and papers cited there). These linear channel properties are but a small subset of all channel properties. They are characterized by many less parameters than the full scale behavior of the channel over a wide range of voltage.

For these reasons I emphasize the analysis of the linear properties of cells and tissues. This linear analysis must, of course, be extended to include the full nonlinear properties if we are to confront much of biological interest. However, the

nonlinear analysis cannot be seriously undertaken until the
structural analysis is complete, until the role of extracellular
space and membrane capacitance is established.

Membrane models. The analysis of linear properties requires
a model of a small piece of each homogeneous membrane.
Historically, the membrane was usually represented as an ideal
resistor in parallel with an ideal capacitor, although more
complex models are sometimes used (see the article by Mathias in
this volume). With the discovery that a large component of
membrane capacitance is markedly nonlinear, varying with both
voltage and time (Schneider and Chandler, 1973; Armstrong and
Bezanilla, 1974), this assumption is no longer self-evident;
indeed, it is obviously unjustified! Fortunately, analyses in the
literature (assuming linearity) can still be used, if experiments
are done in a narrow range of voltages. The full frequency
dependent properties of the membrane capacitance can be included
in the structural analysis if an empirical expression for the
properties of the capacitance is known. This expression can then
be substituted for the appropriate capacitance variable in the
expressions already derived.

The voltage dependent properties of the membrane capacitance
can be easily accommodated only if experiments are done over a
narrow range of voltages, in which an accurate expression for the
frequency dependent properties is known. If the experiments are
done over a full range of voltage, the general applicability of a
circuit approach is not clear. Some circuit modelling of the
nonlinear capacitance may be useful (see, for example, the
appendix of Mathias et al., 1980, where a circuit model is used to
describe the properties of nonlinear capacitance of skeletal
muscle fibers), but one must be skeptical of the general utility

of the approach. In the case of steep voltage dependence, one probably must resort to full scale simulations.

The complications caused by the frequency and voltage dependence of membrane capacitance do not change the goals and need for structural analysis, but the strategy and tactics are changed, and the execution of the analysis is made more difficult. At the present time, for example, such a structural analysis, including the voltage and frequency dependence of membrane capacitance has not been completed, even for the simplest preparation, squid axon, because the best measurements there (Bezanilla et al., 1982; Fernandez et al., 1982) still depend on the series resistance in a complex manner not fully understood experimentally or structurally.

Despite these reservations, we now proceed with a traditional linear structural analysis, following the time honored human behavior of activism; it is better to proceed imperfectly, than to sit paralyzed, waiting for perfection.

Field equations and cable theory. Using the assumption of linearity, structural theory derives field equations (which are a combination of conservation laws and Ohm's law) and a boundary condition describing each homogeneous membrane. These field equations are of varying degrees of complexity, depending on the problem at hand. Jack et al., (1975) provides the best physiological introduction to the one dimensional case -- traditional cable theory. Three dimensional field equations in single cells, or a tissue which can be approximated as a single cell, are described in Eisenberg and Johnson (1970), Adrian et al. (1969b), Peskoff and Eisenberg (1973, 1975), and Peskoff et al. (1976).

Three dimensional field equations for syncytial tissues,
consisting of many cells electrically coupled one to another, are
described in Eisenberg et al. (1979), where a historical
discussion of earlier work with an intracellular perspective is
presented. Syncytial tissues have been analyzed from an
extracellular point of view by Miller and Geselowitz (1978), and
Plonsey and Rudy (1980) and earlier workers cited in those papers.

Perturbation analysis. The field equations for both the
single cell and syncytial problems appear formidable, particularly
when coupled with the appropriate 'membrane' boundary condition
(derived in Peskoff and Eisenberg, 1975), which is a mixture of
the Dirichlet and Neumann boundary conditions extensively analyzed
in texts of potential theory. Nonetheless, these problems admit
of pleasingly simple solutions if the techniques of singular
perturbation theory are used. Perturbation methods systematically
exploit small parameters which appear in differential equations
and frequently permit drastic simplifications of forbidding
problems. These simplifications correspond to problems of reduced
complexity equivalent to classical problems familiar to
experimental workers. Kevorkian and Cole (1981) is a quite
complete presentation of perturbation techniques and is accessible
to those with only a nodding acquaintance with partial
differential equations. Kevorkian and Cole also includes a
discussion of potential spread in spherical and cylindrical cells.

Perturbation theory is particularly well suited for problems
involving membranes, because a small parameter appears naturally
in the boundary conditions for these field equations. The
membrane resistance is always very large compared to the
resistance of the intracellular or extracellular solution (when
written in comparable units). The biological function of
membranes, their evolutionary reason for being, is to isolate the

interior of cells from the external world. Membranes isolate the
interior of cells by preventing the flow of water and solutes
which might perturb the cell interior. Thus, membranes by their
very nature have high 'impedance' to all natural flows.

The small parameter exploited by singular perturbation theory
is just the mathematical expression of the biological role of
membranes (Peskoff and Eisenberg, 1973). The small parameter of
the membrane boundary condition is the ratio of the 'conductance'
of the membrane to the 'conductance' of the intracellular solution
(both expressed in similar units). This ratio is small because
the membrane always presents a greater impediment to flow than the
cytoplasm.

Another small parameter arises naturally in field problems
describing syncytial tissues. Those tissues usually consist of
two interdigitating media, the intracellular space, which occupies
the great majority of the tissue volume, and the extracellular
space, which occupies a tiny fraction of the volume. The ratio of
these two volumes is then a small parameter which can be exploited
in perturbation expansions.

Circuit diagrams of tissues and cells. A successful
perturbation analysis of biological field equations produces a set
of reduced equations which usually are identical to the equations
describing a circuit model of the tissue or cell. This circuit
diagram resembles the circuit diagram one might draw directly from
the topology of the tissue or cell, without the intervening
theoretical trek. Indeed, one's first impression may be that
perturbation theory has provided small dividends, given the
investment of effort. But first impressions can be misleading --
the historical fact is that few tissues have been so clearly
perceived by investigators as skeletal muscle was by Falk and

Fatt, 1964. They did in fact deduce the correct circuit
(including point source effects and correction for capacitive
artifact) without theoretical analysis. Even in that case,
however, the morphometric parameters were not correctly included.
It took much later work to describe the effects of branching
(Adrian et al., 1969a; Schneider, 1970) and wiggling of the
tubules (Mathias, 1975; Mathias et al., 1977; Eisenberg et al.,
1977; Mathias, 1983) and it is still not certain that the random
branching network of tubules has been completely analyzed or
described.

In the case of syncytial tissues, current flow in two
interdigitating media, including flow across the membranes which
bound the tissue, was incompletely analyzed many times (see
historical discussion in Eisenberg et al., 1979) before
perturbation expansions were used to resolve the issue.

The reasons for the difficulty in drawing an equivalent
circuit, without perturbation analysis, is seen if one looks at
the mathematical results of the perturbation analysis in all but
the simplest cases. Perturbation analysis was introduced into
applied mathematics because most problems were too complex to
allow successful approximation without a systematic formalized
procedure. It is possible to guess simplified forms of complex
problems, and sometimes correction factors for those
simplifications, by intuitive methods involving systematic
approximation. But it is difficult to avoid logical
inconsistencies; it is difficult to include all correction terms
of equivalent size; and it is particularly difficult to convince
skeptical colleagues of the validity of such approximations. For
these reasons, as well as for reasons of rigor, the methods of
perturbation theory are needed in biological problems.

Thus, the drawing of a correct circuit representation of a tissue seems to require a theoretical analysis, starting with field equations, exploiting small parameters with singular perturbation theory. Only then, after this tedious but necessary process can a simplified representation of the tissue be convincingly derived.

Measurements of electrical properties. Once the circuit diagram is available to describe the electrical properties of the tissue or cell, sensible experiments can be designed to check the analysis. If the analysis survives experimental check, the linear properties of the different components of the tissue can then be measured. These measurements will at first be made under one set of conditions; indeed, at the present time surprisingly little work has been done under multiple conditions. We will return to this subject later and argue that measurements under multiple conditions are an essential component of structrual analysis and must be performed, if the approach is to be fully exploited. But first things first.

Two classes of measurements have been widely used, measurements either in the time or frequency domain. Measurements in the time domain have obvious advantages. 1) The required equipment is ordinarily available in the electrophysiologist's laboratory. 2) They appear to require no specialized knowledge of circuit theory. 3) The methods can be directly generalized to the non-linear time varying properties which are the biological functions of greatest interest. These complex properties can be recognized and dealt with in traditional and well precedented (hopefully correct) ways: the presence of non-linearities need not corrupt the rest of the analysis.

There are also disadvantages to time domain measurements.
1) Theory in the time domain is in fact often quite difficult,
invariably leading to expressions considerably more involved
(i.e., convolved) than those in the frequency domain. The output
of most systems can only be described by convolution integrals
difficult and expensive to compute, and harder to understand, in a
qualitative or physical way. 2) Time domain data is notoriously
insensitive to different topologies or circuit values in the
underlying circuits. For both reasons parameters of individual
subsystems (membranes or regions of extracellular space) are often
hard to determine uniquely from time domain data. While the last
sentences describe widely known 'facts' (see citations in
Eisenberg and Mathias, 1980), I realize that the facts may turn
out to depend on unconscious assumptions. At the present time, it
is not known whether the difficulties in determining circuits and
circuit parameters from transient responses are inherent to the
mathematics of time domain analysis, or could be remedied by
changes in the numerical procedures of curve fitting. Resolution
of this problem is eagerly awaited.

Frequency domain methods have advantages and disadvantages
which are almost the inverse of those of the time domain. The
advantages are: 1) The theory involved is much simpler since
convolution integrals are replaced by multiplication or division.
2) Circuit topology and parameters are determined far more
precisely for data of (apparently) equivalent accuracy. The
disadvantages are: 1) Specialized equipment is needed. 2) The
theory, while relatively simple, nonetheless uses complex
variables which are not familiar to experimental biologists. 3)
The inclusion of time varying and voltage dependent processes
requires considerable analysis (see paper of Mathias in this
volume) and it is not yet clear how useful or intuitive the
frequency domain will be in this case.

A potentially useful method of analyzing time domain data can avoid some of the difficulties of curve fitting. If time domain responses are integrated, they can be directly related to linear properties as usually measured in the frequency domain. This method holds much appeal but it is not yet clear whether improved methods of fitting the transient response will not remove that appeal.

Location of measurements. We now turn to the question of where measurements are made. It is clear that measurements made at just one location, of the 'input' impedance, as it is often called, have inherent constraints. Indeed, many circuits and structures of physiological interest have membrane properties which are in principle unmeasurable when observed from just one location under one set of conditions. Membrane properties cannot be measured in those cases for the same reason that the properties of two resistors in parallel (or series) cannot be measured from measurements at a pair of terminals made under one set of conditions: there are more parameters than there are independent observable quantities.

The way to break this constraint is to take measurements at different locations or under different conditions. If the topology, and perhaps some of the circuit parameters, can be assumed to be unperturbed by changing these conditions (or the location of measurements), a great deal of useful information can be derived. For example, Valdiosera et al. (1974a,b) used this method to check models of current flow in the t system and sarcoplasm of muscle fibers. Mathias et al. (1981) used this method to measure the radial variation of internal resistance in the lens. And related techniques have been used in analyses of epithelia to separate the properties of apical and basolateral membranes.

A particularly useful intervention, which can often be arranged to perturb the properties of the tissue in a known way, is to decrease the conductivity of the bathing solution by replacing sodium ions with osmotically equivalent amounts of an impermeant solute, like sucrose. This procedure should have little effect on membrane potential, no effect on membrane capacitance, and a linear effect on extracellular conductivity. (That is to say, the conductivity of extracellular spaces should be proportional to that of the bathing solution). Changes in extracellular sodium concentration should have minimal effects on membrane conductance -- none, if the membrane is impermeable to sodium and the concomitant change in ionic strength is not too severe.

If measurements can be made on one preparation bathed in solutions of different conductivity, a great deal of additional information should become available. For example, with this approach it should be possible to tell if the complex properties ascribed to the membrane capacitance of squid axon are in fact properties of the membrane or might be properties of an impedance in series with the membrane, arising in unanalyzed structural complexity.

Measurements at different locations are certainly required if systems of considerable complexity are to be studied. Consider, for example, epithelia, consisting of a series combination of membranes. It is hard to believe that enough information will ever be available from transepithelial measurements to determine individual membrane properties (Clausen et al., 1979). Rather, one must place an electrode within the epithelial cells, and perhaps within the lateral intercellular space between cells, if one is to measure all physiologically interesting parameters independently (Schifferdecker and Frömter, 1978).

Structural analysis of neurons. An equivalent problem arises
in the analysis of dendritic trees of neurons. Here the cell body
is so remote from the dendrites in which interesting integration
(i.e., spread of current and potential) occurs that measurements
of just input impedance are unlikely to allow unique analysis.
Rather, one must seek measurements in the periphery of the
dendritic tree as well as in the cell soma. Such measurements can
in principle be made by microelectrode penetration, or localized
breakdown of dendritic membrane held to the microelectrode by a
seal with gigohms of resistance. It seems more likely, however,
that optical methods of measuring potential, which would (in
principle) allow measurements anywhere in the dendritic tree, will
be needed to deal with this problem.

Despite many difficulties, and many anatomical, theoretical,
and electrical unknowns, I am confident that all the techniques
are already developed to perform a structural analysis of the
spread of potential in the dendrites of neurons, particularly
neurons in hippocampal slices. Such an analysis will probably be
the most significant application of the structural analysis
advocated here, because an analysis of spread of potential in
dendrites is an analysis of the mechanism of integration in
neurons. In other words, structural analysis of dendritic trees
is a mechanistic analysis of decision making in the nervous
system.

The task will not be simple, however, because the integration
of so many techniques into one project remains a combination of
motivation, skills, and resources not easy to put together.
Furthermore, one must anticipate that interesting aspects of
dendritic integration will be mediated by the specific location of
channels with specific properties within the dendritic tree.
Thus, nonlinear properties will come to the forefront. Certainly,

these cannot be investigated without a previous structural
analysis of the underlying linear properties. But it is equally
certain that investigation will require direct measurements of
channel properties and location (with patch electrode techniques,
labelling with monoclonal antibodies, or labelling with specific
toxins and pharmacological agents) and direct confrontation of
nonlinear simulations with nonlinear experimental measurements.

Summary. This part of this paper has described a systematic
approach to the electrical properties of tissues and cells of
complex structure. Obviously, this approach has weaknesses as
well as strengths. Ad hoc methods (suitable for specific
measurements in specific tissues) are often more efficient. Our
treatment of complex membrane capacitance and nonlinear membrane
properties is obviously incomplete enough to be called
unsatisfactory.

Nonetheless, the questions posed by structural analysis
remain among the most significant in physiology, at least in my
view. We cannot expect to know how a tissue or cell works unless
we can synthesize its natural function using the properties of its
molecular components. Surely we need to know the properties of
the channels which are the most significant molecular components;
surely we must also know the arrangement of the channels and
membranes which convert the flow of current through one channel
into the biological signals and flows of direct functional
significance.

Electrical role of restricted extracellular space. The
structural analysis of electrical properties is in large measure
an analysis of restricted extracellular spaces near and within
cells and tissues. These spaces provide the series resistance
which isolates one part of a membrane from another, making them

electrically and (perhaps) functionally distinct. The widespread
presence of these spaces in so many tissues suggests a common
theme to their existence. Do they mediate a common function in
all the cells in which they are found? Are restricted
extracellular spaces an evolutionary adaptation with a universal
function?

To some extent the answers to these questions are obvious, to
some extent, obscure. Obviously, restricted extracellular spaces
are formed by membranes; thus one obvious general function of
these structures is to increase drastically the area of membrane
and the number of channels imbedded in the membrane. The
significance of this adaptation should not be minimized. It may
in fact have been the original evolutionary reason for the
existence of restricted extracellular spaces.

In excitable cells, particularly skeletal muscle and the
conduction systems of the heart (i.e., sheep Purkinje strands),
tubules or clefts serve another purpose directly involved in one
of the main functions of the tissue, the conduction of the action
potential. In both of those tissues the resistance of the
restricted extracellular space serves to isolate the capacitance
of the cleft membrane from the surface membrane, particularly when
the voltage is changing rapidly, as during the foot and upstroke
of the action potential. Because of the extracellular resistance,
the sodium current flowing across the surface membrane need
depolarize only a small fraction of the total membrane area of the
cells. The sodium current (which is limited by the number of
sodium channels) is thus able to depolarize the surface membrane
much more rapidly than otherwise. Furthermore, at any given time
the current can flow a further distance longitudinally down the
fiber. For these reasons, the conduction velocity of the action
potential in skeletal muscle fibers or Purkinje strands of the

sheep heart is much greater than it would be without the
resistance of the restricted extracellular spaces. Of course,
this argument assumes the prior existence of the extracellular
space. If those spaces did not exist at all, the conduction
velocity would be still higher. Thus, the resistance of the
extracellular space serves a useful function given their
existence, but their existence does not seem to be justified by
the resistance itself.

In the case of skeletal muscle fibers, the restricted
extracellular space exists for an obvious purpose. It is a
necessary property of the tubular system which itself is an
essential link in the chain of mechanisms called excitation
contraction coupling. This subject is well reviewed in Costantin
(1975) and in a forthcoming volume of the Handbook of Physiology
(Peachey and Adrian, 1984).

In the case of cardiac muscle, the purpose of the restricted
extracellular space is not so obvious. Excitation contraction
coupling does not seem to require its existence, so the arguments
used for skeletal muscle fail in this tissue. It seems quite
possible that the extracellular space of cardiac muscle is a
necessary consequence of the syncytial nature of the tissue.
Cardiac muscle consists of many cardiac myocytes coupled together,
which have not fused early in development, as the myocytes of
skeletal muscle do early in embryonic life. The reason for the
lack of fusion is a most interesting question, perhaps arising
from the need for specialized conduction pathways or from the need
of the heart to function even after some cells are injured or die
(as suggested to me by Brenda Eisenberg). In any case the absence
of fusion immediately implies the existence of extracellular
spaces of the type discussed here.

The electrical role of restricted extracellular spaces in other tissues is not so clear; and so we now turn to another class of physiological phenomena, involving both diffusion and current flow, seeking a general significance to the role of restricted extracellular spaces. We will find the electrodiffusion in these spaces has profound consequences which may be central to the functioning of many tissues, skeletal muscle, cardiac muscle, and epithelia.

Electrodiffusion. The analysis of the action potential by Hodgkin and Huxley, which is the paradigm for so much later work in electrophysiology, exploits the difference in time scale between conductance change in the squid axon membrane and the concentration change in its environs. Speaking loosely, the conductance changes, which control and are controlled by membrane potential, take 1-2 milliseconds, whereas the concentration changes which accompany the flow of ions through the membrane take 2-10 milliseconds before they are of comparable effect. The separation of these two time scales is essential if membrane properties are to be analyzed with the assumption of a constant gradient of chemical potential, that is to say, with the assumption that the concentrations of permeant ions on each side of the membrane are reasonably constant.

The analysis of conductance changes (i.e., channel properties in more modern languages) is much more difficult if the time scale of concentration change overlaps the time scale of conductance change. In that case, experiments and analysis must be designed to measure and understand both the concentration change itself and the consequent effects on channel properties.

Necessity and size of concentration change. It is important to remember that concentration changes must accompany current flow

carried by ions. The only questions are "How big is the
concentration change?", "What is its time scale?", and "Are the
size and time scale relevant for natural function or biological
experimentation?"

A simple general expression for the _initial_ rate of change of
concentration is given in Levis et al. (1983):

$$\left.\frac{d[X]}{dt}\right|_{t\to 0} \cong 10^3 \quad \frac{V_F}{V_c}\frac{S_c}{V_F}\frac{g_X}{nF}(U-E_X)$$

where the permeant ion X, of charge n, is at concentration [X],
with equilibrium potential E_X. The volume of extracellular space
is V_c in a tissue volume of V_F. The membrane lining that space
has area S_c with a transmembrane potential of U. The conductance
of 1 cm^2 of the membrane is g_X; F is the Faraday constant (96,500
coulombs/mole); and the factor 10^3 reconciles chemical units of
moles/liter and spatial units of cm. Typical values (for calcium
in the clefts of sheep Purkinje strands) would be $g_X \cong 100$ $\mu S/cm^2$;
$U - E_X \cong 100$ mV; and morphometric parameters giving a cleft width
$w \cong 30$ nm, where the cleft width $w = 2(V_cV_F)/(S_cV_F)$. The
resulting initial rate of change of calcium concentration would be
~33 (mmole/liter) per second.

This approximate calculation immediately shows the
significance of changes in concentration of permeant ions, in the
presence of small extracellular spaces. Extensive numerical
simulations (see, for example, Levis et al., 1983) show that the
initial rate of change of concentration is maintained for a
reasonable period of time (10–100 msec in that case); thus, the
transmembrane flow of ions has a drastic effect on concentration,
at least in the case just cited.

This is not the place to summarize the literature on ion concentration changes (some citations are in Levis et al., 1983); nor is it the place to argue the significance of these changes for each tissue or functional phenomena of interest. Rather, the above calculation should serve as a warning that the assumption of constancy of ion concentrations is inherently implausible and must be justified for each current flow across each membrane.

The implications for interpretation of typical voltage clamp experiments are striking. Interpretations assuming constant concentration gradients are likely to be misleading, certainly quantitatively, and probably qualitatively. Considerable effort must be taken to design experiments to control and/or measure the concentration of permeant ions. Theoretical extrapolations from idealized data to realistic situations must include the possibility of concentration changes.

Functional implications of concentration changes in extracellular space. The implications for tissue function are as great as the implications for the analysis of experiments. It seems inherently unlikely that the tissue will share the view of traditional physiology that such concentration changes are an artifact, an unavoidable side effect of important channel properties. It seems far more likely that such large changes in concentration will be used by the tissue for its own purposes, to modulate or even control its natural function.

Consider, for example, the case of calcium ion in the clefts or tubules of muscle fibers. Concentration changes of the size just derived would have profound effects on the conductances of all ions of interest, since calcium concentration has been found to modulate almost all the channels studied to date. Thus, changes in calcium concentration, and the properties of the

extracellular space, would modulate the channel properties
themselves.

The concentration of ions in clefts and tubules would also be
modulated by 'active' transport systems, deriving energy from ATP
(i.e., the Na/K pump) or from concentration gradients of ions
(i.e., the Ca/Na exchange system). In tissues which are active a
good fraction of the time, one must expect the fluxes mediated by
active transport to be comparable to those through passive
channels (otherwise the membrane could not be in steady state for
the biological lifetime of the cell, tissue, or organism). Only
when the duty cycle of the membrane is very low can one expect the
active fluxes to be small.

If the active transport mechanisms of the tubule or cleft
membrane are as important as I suspect (particularly the transport
mechanisms for calcium), a number of properties of cardiac muscle
and skeletal muscle can be viewed in a new perspective. For
example,

(1) The steep temperature dependence of the shape of the
cardiac action potential might well be a result of changes in the
pump rate for calcium and thus of the calcium concentration in the
extracellular space just outside the cardiac membrane.

(2) The dependence of the duration of the cardiac action
potential on the frequency and pattern of stimulation might well
reflect the variation in calcium (and perhaps potassium)
concentration in the clefts.

(3) The dependence of the size of a contraction in heart
muscle on the frequency and pattern of excitation (even when the
voltage excitation is controlled to be constant in amplitude and
duration) may reflect variations in the calcium concentration in
the clefts and thus the calcium current through the putative

'contraction' channel linking surface membrane to the calcium induced calcium release mechanism of the sarcoplasmic reticulum.

(4) The physiological fatigue of skeletal muscle fibers (Brenda Eisenberg and Gilai, 1979) may be a consequence of changes in concentration in the tubules of skeletal muscle. (I call the response to low frequency stimulation 'physiological' fatigue to distinguish it from the response to tetanic stimulation which involves gross morphological changes.)

(5) The active transport systems of the tubular and cleft membrane may have a dominant role in determining the volume of that extracellular compartment, just as active transport mechanisms determine the volume of intracellular compartments. In that case, a long literature describing volume changes in tubules of frog skeletal muscle (and clefts of cardiac muscle as well) needs reinterpretation and re-experimentation using blockers of active transport, like low temperature.

(6) The apparent use dependence of many drugs, particularly calcium blocking agents, may reflect the variation of local calcium concentration in the clefts surrounding the calcium 'channels'. In this situation the use dependence would be expected to vary with pump activity, whether that is varied by temperature, pharmacological blockers, or other experimental or natural agents.

(7) Some of the apparent diversity in properties of calcium channels, as reported by different investigators, particularly with respect to their mechanism of inactivation, may reflect different pump rates and thus local calcium concentrations under different experimental conditions.

(8) Some of the diversity of calcium channel types, found in one laboratory under one set of conditions, may reflect the local environment of the channel. A channel sitting next to a calcium pump would have quite different properties from the same channel

isolated in a membrane, particularly if the channel and pump were
under the umbrella of a calcium binding site.

(9) The variation of calcium concentration in the clefts or
tubules may signal a variety of membrane bound enzymes to create
or modify intracellular messengers (often indirectly through a
cascade of enzymes) which in turn modify the contractile
apparatus, determine the isozymes of the various contractile
proteins, and control cell metabolism in general.

(10) The influence of active transport mechanisms on local
calcium concentrations may complicate the interpretation of
experiments in bathing solutions of reduced calcium concentration
(as suggested by Miyamoto and Racker, 1982; see discussion and
references in Eisenberg et al., 1983). The combination of pump
activity and the existence of a calcium binding protein might make
it difficult if not impossible to exhaust the supply of calcium
available to a 'contraction channel', particularly if the pump and
binding protein are in close conjunction to the channel, supposing
such exists at all.

(11) Finally, one should point out an ironical consequence of
these effects, if they are as significant as I expect. Consider
cells isolated from syncytial tissues by proteolytic enzymes, for
example, cardiac cells. The shape of the action potential of
these cells is normally taken as an indicator of their viability;
if the shape is similar to that in intact tissues, the cells are
thought to be normal. If changes in cleft concentration are
significant modulators of the action potential, the shape of the
action potential should be different in isolated cells, even if
the membrane properties of the cells are the same as those in the
natural syncytial preparation. In fact, it seems that the action
potentials could be normal only if the extracellular concentration
did not modulate the action potential or if active transport
mechanisms were potent enough to maintain the concentration
constant outside the membrane. In the latter case, of course,

temperatures at which active transport is sluggish or inoperative, should have a substantial effect.

It is possible, of course, that changes in the concentration in extracellular space are not as large as one would expect. They may be buffered or controlled by negative feedback, in some way. But that would not necessarily decrease the role of extracellular spaces in the phenomena just described.

Consider, for example, the effects of a membrane active transport system on ions in narrow extracellular spaces, if the active transport system has evolved to control the extracellular concentration of ions, the concentration of ions in the cleft. This mechanism would homeostatically control the concentration of ions and the volume of an extracellular compartment, just as other active transport mechanisms homeostatically control the concentration and volume of the intracellular space. The ion concentration in the extracellular space would remain quite constant, during physiological activity, but the rate of active ion transport would vary drastically, as it produces the membrane flux needed to keep that concentration constant. Since the membrane enzyme responsible for active transport has intracellular substrates and products (as well as the extracellular product of ion flux), important changes would occur within the cell. Thus, the cell would have converted a putative extracellular concentration change into an intracellular signal. It would control the extracellular concentration of an ion by varying active transport of the ion across a membrane, just as an operational amplifier controls its input voltage (to a virtual ground) by varying its output voltage and current. The output of an operational amplifier is a useful (i.e., amplified and buffered) measure of the input, even though the input is virtual, controlled by negative feedback to a negligible size. Similarly, the cell would have many intracellular outputs it could use as intracellular signals of the (virtual) change in extracellular concentration. The homeostatic active transport mechanisms would provide negative feedback to make the concentration change in the clefts virtual; the active transport mechanisms would do this by controlling the transmembrane fluxes of the transported ions. As a consequence, the local intracellular concentration of transported ions would change and so would the local intracellular concentration of other substrates of the membrane transport process. In this way the cell could simultaneously buffer the concentration of an important ion in the extracellular space and derive an important intracellular signal 'proportional' (or, more accurately, a monotonic function of) the integrated flux of that ion.

Experimental evaluation of extracellular concentrations. The role of extracellular cleft concentrations in physiological function can be directly evaluated with a (relatively) small extension of present day techniques. Indicator dyes developed to measure (for example) intracellular calcium concentration might be applied extracellularly and the resulting optical absorbance could be calibrated and used to measure extracellular concentration. In the case of muscle fibers, fluorescence signals might be particularly useful in easing the calibration problem, in separating signals from the tubules or clefts from signals in the bath. It might be wise to adapt the experimental set-up of Blinks (1965) in which the exciting light is applied in a plane perpendicular to the fiber axis, and the fiber is observed looking down on the plane of the exciting light. Blinks photographed the cross sectional shape of fibers held vertically below a long working distance objective. He looked down the axis of the fiber and directly observed its illuminated cross section. The light he observed was scattered at 90°. The only illumination present was

a thin plane of light at right angles to the fiber axis. This set
up could be used to observe a fiber bathed in a solution
containing an impermeant calcium indicator, if the thin plane of
light were at the wavelength to excite fluorescence of the
indicator and the observation were made at the emission
wavelength. A fluorescent signal from a dye in the extracellular
space within the fiber would then be visible with little
contamination from dye in the bathing solution outside the fiber.

Specific experimental design can also be used to isolate
signals arising within clefts or tubules from signals in the
bathing medium. The experimental design can be self-calibrating
and itself provide some of the necessary controls. Consider, for
example, experiments using temperature to modulate the activity of
an active transport system in the cleft or tubule membrane.
Measurements made in the cold are likely to reflect just passive
properties, because active transport systems are unlikely to
function well at those temperatures. Thus, comparison of dye
signals in the warm and the cold should show if the active
transport systems are able to significantly modulate extracellular
concentrations. The most obvious experiment would compare a
putative tubule/cleft calcium signal in the warm and cold, while
the fiber was bathed in a low calcium (approximatley 200 $\mu\underline{M}$), high
magnesium (5 $m\underline{M}$) solution.

Measurements can also be made exploiting the greater spatial
(but worse time) resolution of the electron beam microprobe. Here
rapid freezing should allow direct visualization of the
concentration of ions in the clefts or tubules under different
conditions.

Extracellular spaces in other tissues. The discussion of the
role of extracellular spaces in other tissues, particularly

epithelia, takes me too far beyond my area of competence. Suffice
it to say, that the lateral intracellular spaces of epithelia are
important, perhaps dominating, determinants of some of their
functions, particularly those involving water flow. The problems
of concentration change in these clefts are classical problems of
epithelial physiology and have been a significant theme of
research in this field for many years. Workers, like myself,
interested originally in excitable tissues are the newcomers to
these problems.

Extracellular spaces as an evolutionary adaptation. These
speculations suggest that restricted extracellular spaces may have
a unifying rationale. They may be a universal adaptation to allow
a cell to measure, use or control the change in extracellular
concentration, which is an integral part of the membrane flux and
thus cell activity.

Like all speculations this certainly is well beyond the reach
of knowledge. Like all speculations, it requires much thought and
testing before being taken too seriously. Nonetheless, unlike
some other speculations (particularly in the next part of this
paper), these are directly testable. One can hope that such tests
will be fruitful even if the hypotheses themselves are refuted.

PART 2:   BIOLOGICAL INVESTIGATION: THE UNDERSTANDING OF
          EVOLUTIONARY ADAPTATIONS

Introduction
     The second part of this paper tries to view biological
investigation rather generally, looking at all biological systems
and structures as a result of the same evolutionary mechanisms,
operating at different length scales. In this view, investigation

of the functional consequences of the structure of hemoglobin must
follow the same strategy and tactics as the investigation of the
functional consequences of the structure of the knee joint,
however different the vocabulary of chemistry is from that of
gross anatomy!

Until recently the study of biology meant the study of
objects larger than the wavelength of light. With that
resolution, so limited as we now know, an astonishing amount was
learned about principles which govern the structure, function,
natural history, and evolution of living organisms. The
significant questions of biology were understood. Biologists
learned to ask "How are animals and plants built?", "How do they
function?", "How do they live together?", and "How did they
evolve?" Scientists working on each of these questions clustered
into academic disciplines: anatomy, physiology, ecology, and
paleontology.

Central Role of Evolution. These academic disciplines
recognized the central role of evolution in each biological
question. Analysis of structure reveals magnificent adaptations,
as we all admire in the intricacies of the anatomy of muscles, of
the nervous system, or of sense organs. But analysis of structure
also illustrates missed chances, muddles, and misadaptations of
evolution, evidently the result of the compulsive opportunism of
evolutionary process and the constraints imposed by the
environment and the raw materials for adaptation. Examples of
misadaptation are numerous. The inappropriate location of retinal
neurons in the light path of vertebrate eyes contrasts with their
appropriate location in cephalopod eyes. In mammals, breathing in
and out through the same pipe guarantees inefficient mixing of
oxygen rich 'fuel' and carbon dioxide rich waste. Oxygen exchange

in fish is far more efficient, because of the morphological
separation of oxygen rich and oxygen poor water.

Analysis of function reveals the same traces of its
evolutionary designer: the choices open to evolution were clearly
limited. No wheels or metallic wires were found in animals, and
even the magnificent expediency of evolution could not overcome
the resulting constraints on function. Wheels and wires would
certainly improve the design, the fitness of many biological
systems.

The last three decades have brought a remarkable increase in
the resolution with which we study life. Molecular biologists
have developed tools to study the molecules which perform so many
living functions. Indeed, it seems that surprisingly many of the
adaptations of evolution have occurred on the molecular scale, the
familiar story of 'the chemical basis of inheritance' being the
prime example. Macroscopic structures are of little importance in
the mechanism of heredity in prokaryotes; they are of limited
importance (as far as we now know) in eukaryotes. Along the same
lines, we have learned that the intramolecular properties of
enzymes and proteins (e.g., their tertiary structure) contain many
of the adaptations which define life. Much of metabolism requires
little macroscopic organization of enzymes: soluble enzymes
reproduce complex metabolic cycles whether uniformly dispersed in
test tubes, or cells.

Role of structure on different length scales. Mechanisms
depending on macroscopic structure are important even in classical
biochemistry. The mechanisms which produce ATP from oxygen
involve evolutionary adaptations in both molecular and
supramolecular structures. Once solubilized, mitochondrial
enzymes cannot produce ATP from oxygen. These enzymes must be

organized into membranes across which a voltage is maintained and through which current flows in specific ionic channels. Oxidative phosphorylation was understood only when the role of the mitochondrial membrane was considered, when workers confronted the potentials across and currents through the mitochondrial membranes, as well as the enzymes, reaction pathways, and putative high energy intermediates which suffice to describe many other metabolic pathways. Structures on the molecular and atomic length scale are intimately involved in oxidative phosphorylation, but so are structures on the much larger length scale of membranes.

The study of membrane processes is now exploiting an incredible increase in the resolution of experimental technique. After all, it was only thirty years ago that the role of membranes in neurophysiology was properly perceived. Today, as this book shows better than most, questions are asked (and answered) about individual molecular channels which govern the flux of molecules and current through biological membranes. Progress in this field will undoubtedly open new vistas equivalent to those revealed to molecular biologists in the last 30 years.

But increased resolution can obscure processes, and mislead us, as well as enlighten us. It is hard to find a pass through a mountain range when one moves on foot, observing structures of meters and kilometers in size. It is easy to find the pass through the mountains looking from an airplane or satellite, observing a larger area with a lower resolution. Of course, neither approach works by itself. The pass seen from the airplane may prove to be blocked by 'invisible' boulders a few meters in diameter; and the pass over the next ridge may be hidden when walking in a steep valley. In biology, as in map making, observations are needed at all relevant resolutions, asking

questions at each scale of time and distance, if one is to find
the resolution at which the relevant answer is apparent.

Resolution of experiments must match the length scale of
adaptations. The need to observe and question at all resolutions
is particularly important when studying the outcome of evolution.
Sometimes evolutionary adaptations have occurred on the scale of
atoms, as we see in the magnificent hemoglobin molecule which
transports oxygen so well in the blood (Perutz, 1978; Fermi and
Perutz, 1981). Sometimes adaptations occur on a much larger
level, as we see in the shape of fish or sea mammals. Often
adaptations occur on multiple scales of distance: the movement of
limbs and animals is importantly influenced by the gross
morphology and tendon insertions of whole muscles, as well as by
intramolecular adaptations of the thin filaments of an individual
sarcomere.

The study of membranes is one of those subjects that requires
analysis on different scales of time and distance. As we have
seen in such beautiful detail in this book, membrane adaptations
occur on the molecular scale. It is truly exciting that we can
now study these molecule by molecule, one at a time, with the
molecules in their natural location, performing their natural
function.

Membrane adaptations on different length scales. But
adaptations also occur on the length scale of membrane
architecture; they do not solely occur in and within channels.
The distribution of channels in membranes are likely to be as
important as the properties of channels themselves. We expect
that some evolutionary problems were solved by the selective
placement of channels in different membranes, as well as by
adaptations within the channels themselves. Many membranes are

folded and invaginated, making the clefts and tubules which are
such prominent anatomical features of cells and tissues. We must
presume that this membrane architecture is functionally important,
just as we must presume that the placement of channels is
important, even if we do not yet understand the functional role or
evolutionary heritage of these structural complexities.

The location of channels is particularly important because
different membranes, or different parts of continuous membranes
are functionally differentiated by their location. Thus, the same
channel may have a quite different function depending on its
location in a plasma membrane facing the extracellular world, in a
plasma membrane facing a cleft, or in an organelle membrane,
facing the cytoplasm.

Even the outer plasma membranes are nonuniform; they too are
extensively differentiated to perform their function. Channels
are not randomly located. They are placed where they are needed,
receptors near the source of effectors (e.g., neural
transmitters). Indeed, if we find a channel to be in many types
of outer membranes, we should guess that it serves a universal
function.

Consider, for example, the calcium activated potassium
channel which is found in membranes with widely varied function.
One might guess that this channel protects damaged cells from
prolonged depolarization and consequent swelling and death.
Damage to the cell membrane would be expected to depolarize the
cell, making some cells more permeable to sodium and perhaps
calcium. These ions migrate across the membrane and water flows
with them, producing swelling which, if severe enough, would
rupture the membrane and kill the cell. An increase in the
calcium permeability of the membrane has a counteracting effect,

if the membrane also contains calcium activated potassium
channels. When the internal concentration of calcium rises, these
potassium channels are activated, increasing the specific
conductance of the membrane, holding it close to the equilibrium
potential for potassium, thus preventing the depolarization which
would otherwise have resulted from the membrane damage.

The previous discussion shows that we must study all relevant
length scales as we seek to understand membranes, as well as
living processes. Attention to only one scale, however exciting
or seductive that may be, will not give useful insight into
biological problems that evolution solved on different scales of
organization.

Relevant biological questions are defined by evolutionary
adaptations. We face another danger resulting from the success of
molecular biology and the rush to exploit that success: we can
easily forget our questions.

The questions in biology have always differed from the
questions of physical sciences, just as questions in historical
sciences (archetypes: geology and paleontology) differ from those
in experimental sciences (physics and physiology). The nature of
this difference has been hard to explain or analyze, particularly
to experimental physical scientists, but recent work in the most
analytical of physical sciences, applied mathematics, is perhaps
helpful. A class of equations has been discovered, simple
equations at that, which have remarkable properties, giving
chaotic solutions which appear random, and yet arise from strictly
deterministic systems (May, 1976; May and Oster, 1976; Gurel and
Rossler, 1979; Ruelle, 1980; Feigenbaum, 1980; Guckenheimer,
1982). These chaotic systems are extraordinarily sensitive to the
way the system starts, that is, to the initial conditions. One

can presume that chaotic systems are just as sensitive to external interventions, to environmental perturbations, as they are to their initial conditions. External interventions are a form of re-set, in effect a restart of the system with new initial conditions.

Along with many others, I cannot resist the speculation that such chaotic equations govern many strictly physical processes which have traditionally been considered random and analyzed by the theory of stochastic processes. Statistical mechanics, at least classical statistical mechanics governed by strictly deterministic laws, is one candidate. Einstein might also have tried to cast quantum mechanics itself in such a form, if he had preferred a chaotic to a random universe. Indeed, I suspect important insights will emerge as chaotic systems are analyzed with the traditional tools of stochastic analysis. The insights are likely to be important to both the random and the deterministic theories, just as the interaction of potential theory and theory of stochastic processes has been so mutually beneficial.

Limitations in the analysis of chaotic systems. Imagine that historical and evolutionary processes are governed by equations of such a chaotic class, as I believe they are. Then I would argue that the nature of human understanding is inherently restricted, the class of meaningful questions is proscribed, when we study the outcome of a chaotic historical or evolutionary process.

Inherent limitations in human understanding should not be surprising. Since Gödel, mathematicians have become familiar with unprovable but true propositions, even in strictly deductive systems. Mathematicians recently have realized that proofs of some true and provable theorems may exceed human capacities, if,

for example, they take more than a lifetime to read! Physicists
are familiar with systems which can only be probed in limited ways
-- knowledge of atomic and subatomic processes (and certain
macroscopic systems as well, such as superfluidity or
superconductivity) is limited by the inherently probabilistic laws
of quantum mechanics. Deterministic questioning of quantum
processes has been proven fruitless -- it is not possible, and
thus is wasteful of scientists' time and effort as well as
unsuccessful in providing new information.

Similarly, I think that chaotic processes can be meaningfully
studied in only limited ways, essentially by probing the actual
outcome of the (chaotic) historical process, asking the
existential question "What is?" instead of the metaphysical,
nearly theological, "What might have been?" People, and certainly
governments, do not, and never will, have the energy or resources
to probe every possible outcome of a chaotic process, even if that
were a finite set. They confine themselves to a narrow range of
questions around the central questions: "What is there?", "What
does it do?", "How does it work?", "How did it happen?". We must
refrain from asking too many of the unanswerable questions: "What
might be there?", "How could it work?", "How might it have
happened?". Of course, it is fun to ask these unanswerable
questions. But scarcity of resources must eventually limit the
integral of pleasure, and most of us are constrained by practical
considerations not to spend too much time working on interesting
but inherently unanswerable questions. When studying life, we
should be restricted in what we can ask, at least if we view
organisms and biological systems as the outcome of a chaotic
process. We must avoid questions, far from the natural function
or history of the system, if we are to use our energies
efficiently, proceeding in our task of understanding the systems

before us. Our task is to analyze a particular outcome of a
chaotic evolutionary process.

Biology and a grand inverse problem. The relevant questions
listed previously in this article fall into two classes. The
questions "How is it built", "What does it do?", and "How does it
work?" are quite different from the question "How did it get that
way?". The first two questions are the classical questions of
experimental science and are amenable to the usual approaches.
The historical question is much more difficult because it requires
the reconstruction of the actual outcome of a chaotic historical
evolution. Perhaps that is why experimental biologists often
identify the legitimate historical question "How did it evolve?".
with the teleological question "Why did it evolve"

The evolutionary history of a biological system is hard to
reconstruct for two reasons. The first is the paucity of
historical data and the fact that most of the data we have is from
one time point, namely the present. The second is the essentially
chaotic nature of evolution. Chaotic processes are so sensitive
to external disturbances that one needs an incredible density of
experimental information before one can understand the cause of a
particular event. The smallest unobserved or unknown perturbation
can deflect the entire subsequent historical development.

Both of these difficulties in reconstructing an evolutionary
process are reminiscent of the difficulties mathematicians face
when solving 'inverse' problems even of well defined non-chaotic
systems. The difficulties in solving inverse problems can best be
understood by contrasting a classical inverse and forward problem.
A classical forward problem is the solution of a differential
equation and boundary condition for, say, the temperature
distribution in the earth as a result of a certain distribution of

heat sources. The corresponding inverse problem would be the determination of the distribution of heat sources from a limited set of experimental information, e.g., the temperature on the surface of the earth. Inverse problems have a certain fascination because, like most human thought but unlike traditional mathematics, they rarely have unique solutions. Even in those cases where the solution is theoretically unique, it may not be practically unique. The unique solution of an inverse problem may be computationally unstable, sensitive to tiny perturbations or errors in the data or numerical analysis. The art and essence of solving an inverse problem, like most of experimental science, is in the casting and recasting of the problem until a prescription is found which tells what data should be measured, and how it should be processed, to give a reliable result. Sometimes the only possible reliable results may not relate directly to the original question. There may not be any possible reliable answer to the original question, using the data available. In this case, a well-determined solution of the original problem may require a new type of measurement, or it may be simply impossible, as, for example, in many chaotic systems.

Given these difficulties in solving even simple inverse problems, I think we can see why reconstruction of the history of a biological system is forbidding. The experimental biologist is instinctively correct in avoiding the teleological question, "Why did it evolve that way?"

Relevant questions in molecular biology. What I have written will hardly be surprising to the evolutionary biologist and will be trite to the mathematical analyst of chaotic systems. But the molecular biologist may not be so aware of these constraints on our questioning of biological systems. He will be tempted to study all he can of the magnificent proteins and channels we have

before us. He may not realize that the analysis of <u>all</u> the
properties of these molecular machines is one of the forbidden
questions. No doubt all the properties of these molecules are
fascinating, and there certainly are enough properties to keep
molecular biologists in work for many years. But why study all
these properties? Shouldn't we restrict ourselves to those
relevant to the biological function for which the molecule was
evolved? Cannot we expect these relevant questions to be
specific, and thus easier to answer than the more numerous
questions, mostly concerning chemical and physical properties
irrelevant to biological function or evolution?

In my opinion, the relevant biological questions for
molecules are the same as those for cells and tissues. The
questions are the same on the molecular and macroscopic scales
because the process that built the molecules is the same on both
scales. How is the molecule built? What biological function does
it perform? How does it do it? How did it get that way? These
questions are as central to molecular biology as they are to the
biology of the last 150 years. They differ only in the length
scale at which they are asked and the vocabulary in which they are
answered.

<u>Productive questions in molecular biology</u>. Constraining his
questions to the biologically relevant, the molecular biologist is
likely to reap a welcome reward. He may find it easier to get
answers. Investigation of natural protein structure is much
easier than the investigation of arbitrary polypeptides because
proteins are a tiny subset of the topologically possibly
polypeptide structures. Proteins have evolved as <u>un</u>branched
polypeptides rarely (if ever) tied into knots. Similarly, we can
expect that other properties of proteins <u>relevant to the function</u>
<u>for which they were evolved</u> are a small, even tiny, subset of all

the chemical properties of such complex molecules. Perhaps by
confining his attention to the <u>physiological</u> properties of his
molecules, the molecular biologist will find his work much easier,
as well as biologically more relevant.

Evolutionary processes proceed always blindly, always
opportunistically, usually bluntly, sometimes stupidly, sometimes
subtlely, on a macroscopic scale. It is hard to believe they
proceed differently on the molecular scale. Thus, we should find
microscopically what has been found macroscopically. The
biologically appropriate questions, those probing the natural
function for which the molecule or system has evolved, are much
easier to answer than questions concerning the general properties
of the molecule or system. Properties with little adaptive value
are unlikely to have been selected and simplified by the processes
of evolution.

<u>Relevant and productive questions in membrane biology</u>. We
then have a set of questions to apply to the biological systems,
of whatever size. How are they built? What do they do? How do
they do it? How did they evolve? These questions are the
questions of the classical biological disciplines of anatomy,
physiology, and paleontology. They also seem to me to be the
relevant questions of molecular biology and its nascent
subdisciplines.

ACKNOWLEDGEMENT

My tutor, John Edsall, Professor Emeritus, Harvard
University, to whom this paper is dedicated, taught me biology
from paleontology to x-ray crystallography. Brenda Eisenberg
discussed these ideas and edited the manuscript from her

delightfully pragmatic point-of-view. It is a pleasure to thank
my mentors and collaborators in structural analysis (V. Barcilon,
J. Cole, C. Clausen, L. Costantin, B. Curtis, Brenda Eisenberg,
E. Engel, P. Fatt, G. Falk, P. Gage, P. Horowicz, J. Howell,
E. Johnson, J. Leung, R. Levis, R. McCarthy, R. Mathias,
R. Milton, B. Mobley, A. Peskoff, J. Rae, R. Thomson,
R. Valdiosera, P. Vaughan) who have shared in the scientific work
described here. But the speculations are my responsibility,
particularly those which will prove false, or worse, ignorant.

This work was supported by grants from the NIH (HL 20230) and
Muscular Dystrophy Association.

REFERENCES

Adrian, R. H., Chandler, W. K., and Hodgkin, A. L., 1969a, The
    kinetics of mechanical activation in frog muscle, J.
    Physiol., 204:207.
Adrian, R. H., Costantin, L. L., and Peachey, L. D., 1969b, Radial
    spread of contraction in frog muscle fibers, J. Physiol.,
    204:231.
Armstrong, C. M., and Bezanilla, F., 1974, Charge movement
    associated with the opening and closing of the activation
    gates of the Na channels, J. Gen. Physiol., 63:533.
Bezanilla, F., Taylor, R. E., and Fernandez, J. M., 1982,
    Distribution and kinetics of membrane polarization. I.
    Long-term inactivation of gating current, J. Gen. Physiol.,
    79:21.
Blinks, J. R., 1965, Influence of osmotic strength on cross-
    section and volume of isolated single muscle fibers, J.
    Physiol., 177:42.

Clausen, C., Lewis, S. A., and Diamond, J. M., 1979, Impedance analysis of a tight epithelium using a distributed resistance model, Biophys. J., 26:291.

Costantin, L. L., 1975, Contractile activation in skeletal muscle, Prog. Biophys. Molec. Biol., 29:197.

Diamond, J. M., and Machen, T. E., 1983, Impedance analysis in epithelia and the problem of gastric acid secretion, J. Memb. Biol., 72:17.

Dormer, K. J., 1980, "Fundamental Tissue Geometry for Biologists," Cambridge University Press, New York.

Eisenberg, B. R., and Cohen, I. S., 1983, The ultrastructure of the cardiac Purkinje strand in the dog: a morphometric analysis, Proc. R. Soc. Lond., 217:191.

Eisenberg, B. R., and Gilai, A., 1979, Structural changes in skeletal muscle fibers after stimulation at a low frequency, J. Gen.Physiol., 74:1.

Eisenberg, R. S., Barcilon, V., and Mathias, R. T., 1979, Electrical properties of spherical syncytia, Biophys. J., 25:151.

Eisenberg, R. S., and Johnson, E. A., 1970, Three dimensional electrical field problems in physiology, Prog. Biophys. Mol. Biol., 20:1.

Eisenberg, R. S., and Mathias, R. T., 1980, Structural analysis of electrical properties of cells and tissues, Crit. Rev. Bioengr., 4:203.

Eisenberg, R. S., Mathias, R. T., and Rae, J. L., 1977, Measurement, modelling and analysis of the linear electrical properties of cells, Ann. N.Y. Acad. Sci., 303:342.

Eisenberg, R. S., McCarthy, R. T., and Milton, R. L., 1983, Paralysis of frog skeletal muscle fibers by the calcium antagonist D-600, J. Physiol., 341:495.

Falk, G., and Fatt, P., 1964, Linear electrical properties of striated muscle fibers observed with intracellular electrodes, Proc. R. Soc. Lond. B. Biol. Sci., 160:69.

Feigenbaum, M. J., 1980, Universal behavior in nonlinear systems. Los Alamos Sci., 1:4.

Ferme, G., and Perutz, M. F., 1981, "Haemoglobin and Myoglobin. Atlas of Molecular Structures in Biology." Oxford University Press, London.

Fernandez, J. M., Bezanilla, F., and Taylor, R. E., 1982, Distribution and kinetics of membrane dielectric polarization. II. Frequency domain studies of gating current, J. Gen. Physiol., 79:41.

Guckenheimer, J., 1982, Noise in chaotic systems, Nature, 298:358.

Gurel, O. and Rossler, O. E. (editors), 1979. "Bifurcation Theory and Applications in Scientific Disciplines," Annals N.Y. Acad. Sci., Vol. 316, New York.

Jack, J. J. B., Nobel, D., and Tsien, R. W., 1975, "Electrical Current Flow in Excitable Cells," Clarendon Press, Oxford.

Kevorkian, J. and Cole, J. D., 1981, "Perturbation Methods in Applied Mathematics," Springer-Verlag, New York.

Levis, R. A., Mathias, R. T., and Eisenberg, R. S., 1983, Electrical properties of sheep Purkinje strands. Electrical and chemical potentials in the clefts. Biophys. J., 44:225.

Mathias, R. T., 1975, A study of the electrical properties of the transverse tubular system in skeletal muscle. Ph.D. Dissertation, University of California, Los Angeles.

Mathias, R. T., 1983, Effect of tortuous extracellular pathways on resistance measurements, Biophys. J., 42:55.

Mathias, R. T., Eisenberg, R. S., and Valdiosera, R., 1977, Electrical properties of frog skeletal muscle fibers interpreted with a mesh model of the t-system, Biophys. J., 17:57.

Mathias, R. T., Levis, R. A., and Eisenberg, R. S., 1980, Electrical models of excitation contraction coupling and charge movement in skeletal muscle, J. Gen. Physiol., 76:1.

Mathias, R. T., Rae, J. L., Eisenberg, R. S., 1979, Electrical
    properties of structural components of the crystalline lens,
    Biophys. J., 25:181.

May, R., 1976, Simple mathematical models with very complicated
    dynamics, Nature, 261:459.

May, R. M. and Oster, G. F., 1976, Bifurcations and dynamic
    complexity in simple ecological models, Amer. Naturalist,
    110:573.

Meinhardt, H., 1979, The random character of bifurcations and the
    reproducible process of embryonic development, Annal. N.Y.
    Acad. Sci., 316:188.

Mikkelsen, R. B., and Wallach, D. F. H., 1976, Photoactivated
    cross-linking of proteins within the erythrocyte membrane
    core, J. Biol. Chem., 251:7413.

Miller, W. T., and Geselowitz, D. B., 1978, Simulation studies of
    the electrocardiogram. I.   The Normal Heart, Circ. Res.,
    43:301.

Miyamoto, H., and Racker, E., 1982, Mechanism of calcium release
    from skeletal sarcoplasmic reticulum, J. Memb. Biol., 66:193.

Peachey, L. D., and Adrian, R. (editors) 1984, "Handbook of
    Physiology, Section 10:   Skeletal Muscle," Williams and
    Wilkins, Baltimore.

Peskoff, A., and Eisenberg, R. S., 1973, Interpretation of some
    microelectrode measurements of electrical properties of
    cells, Ann. Rev. Biophys. and Bioeng., 2:56.

Peskoff, A., and Eisenberg, R. S., 1975, The time-dependent
    potential in a spherical cell using matched asymptotic
    expansions, J. Math. Biol., 2:277.

Peskoff, A., Eisenberg, R. S., and Cole, J. D., 1976, Matched
    asymptotic expansions of the Green's function for the
    electrical potential in an infinite cylindrical cell,
    SIAM. J. Appl. Math., 30:222.

Plonsey, R., and Rudy, Y., 1980, Electrocardiogram sources in a
    2-dimensional anisotropic activation model, Med. & Biol.
    Eng. and Comput., 18:87.

Ruelle, D., 1980, Strange attractors, Math. Intell., 2:126.

Schifferdecker, E., and Frömter, E., 1978, The AC impedance of
    Necturus gallbladder epithelium, Pflügers Archiv., 377:125.

Schneider, M. F., 1970, Linear electrical properties of the
    transverse tubules and surface membrane of skeletal muscle
    fibers, J. Gen. Physiol., 56:640.

Schneider, M. F., and Chandler, W. K., 1973, Voltage dependent
    charge movement in skeletal muscle: A possible step in
    excitation-contraction coupling, Nature, 242:244.

Valdiosera, R., Clausen, C., and Eisenberg, R. S., 1974a, Circuit
    models of the passive electrical properties of frog skeletal
    muscle fibers, J. Gen. Physiol., 63:432.

Valdiosera, R., Clausen, C., and Eisenberg, R. S., 1974b,
    Impedance of frog skeletal muscle fibers in various
    solutions, J. Gen. Physiol., 63:460.

INDEX